PUBLISHED JULY 1991

Homebuilders!

They are a special breed! Dedicated. Skilled. Determined. Unafraid to tackle even the most difficult project. Yet however complicated construction becomes, they invariably display a tenaciousness that will not detract them from projected goals.

A homebuilder who exemplifies these characteristics is J.V. Withrow of Central City, KY (left). Withrow was 47 years of age when construction of the Starduster II in background was started. He had absolutely no experience around airplanes — did not know how to fly — had never worked with any of the materials and systems that go into a tube and fabric airplane. He was not deterred, however. "How to" publications were purchased and he proceeded to educate himself with aircraft building techniques. When construction started he sought the help of other EAA members for advice when problems arose. He attended forums and workshops at Oshkosh. Every situation was met head on, and he proceeded only after each component was completed according to his own rigid standards of perfection.

Thirteen years later, at age 60, Withrow's biplane was completed. And what a thing of beauty it was! Workmanship was outstanding. His determination was rewarded and his standards of excellence recognized when the aircraft was flown to the 1985 EAA Oshkosh Convention — piloted by Withrow — where it was named Reserve Grand Champion for a plans built aircraft. A tremendous honor — well deserved. It was just one of many awards in store for this outstanding aircraft and its builder. Withrow's dedication — like that of many other builders in EAA — is just one more example of what aircraft homebuilding is about — this "can-do" attitude that makes the homebuilder SOMEONE SPECIAL!

Right — N2X, a Rutan designed Long-EZ wa built by George Nopper, Tuscon, AZ. After flyir the aircraft for three years and some 300 hours it was sold in 1989 to Rob Eckes. At the 199 EAA convention Nopper and Eckes were awarded Grand Champion honors in the plar built category for this really outstanding hom built. It had a total of 350 hours logged at th time, yet remained in pristine, perfect conditio Engine is a Lycoming 0-235-L2C with a Warnk Almost Constant Speed prop.

PART I

The EAA Aviation Center — fabulous world headquarters for the Experimental Aircraft Association and the EAA Aviation Foundation Museum. Located at Oshkosh, Wisconsin, adjacent to Wittman Airfield, site of the world's largest aviation event — the annual EAA Oshkosh convention.

Our Purpose

Long ago, the need was recognized by the Experimental Aircraft Association for a specifically prepared guide for the amateur aircraft builder. Many changes for the better have followed the amateur built aircraft movement as the result of EAA and its members and the high standards they have set. Standards that have been followed by the amateur builder that have earned the freedoms and opportunity of building and flying your own aircraft plus the many antique, warbird, and unique factory built aircraft operating with Experimental Airworthiness Certificates. We have earned the cooperation and respect of the Federal Aviation Administration as well as the aviation public.

The Experimental Aircraft Association is an organization of members with many talents and interests. It includes aircraft designers, builders and restorers who preserve our aircraft history, including antiques, classics and many of our military aircraft . . . "warbirds". EAA has brought them all together to share this varied interest in aviation.

This Custom Built Aircraft Handbook . . . a guide to construction standards and the FAA rules and requirements for the amateur aircraft builder . . . is in the fourth printing with changes in FAA requirements being brought up-to-date. For requirements in other countries, contact EAA Headquarters.

Aeronautical standards recognize no country's boundary lines. These accomplishments were learned and earned by the many dedicated aviation engineers, hobbyists and pilots since the beginning of flight. It is recommended that the material contained in this handbook be used as a guide in the construction, maintenance, registration and flying of homebuilt aircraft and other such aircraft operating with experimental certificates.

Other than the mandatory applicable FAA FARs, the detailed recommendations contained herein should not be construed as being regulatory or restricting advanced designing or use of newer, and possibly better, materials or construction techniques, but rather as a solid foundation on which to build.

Tom Poberezny
President
EXPERIMENTAL AIRCRAFT ASSOCIATION

Your Responsibilities

(As A Custom Aircraft Builder)

For well over three decades, and since the founding of the Experimental Aircraft Association, the amateur aircraft builders have proven themselves among the most dedicated of aviation enthusiasts. For in this wonderful aeronautical world, seldom do we find people who can design, build and test fly their own creation, and then some share their design with other creative and self-educated aircaft builders.

The amateur aircraft builder of today has earned over these past thirty-three years great freedom as the result of high construction maintenance and flying standards. It is the responsibility of each person who builds his own aircraft to remember that he is not alone . . . his or her very action in building and flying the machine he has created with his own hands affects every other person who is doing the same thing. Poor workmanship or poor flying by a homebuilder in Texas many ultimately take away the privilege of homebuilding from a fellow in Maine. Too many accidents and tragedies combined with political pressures from a non flying public and the right to homebuilding can be revoked, as it was in the past. With a good safety record and ever improving aircraft, we homebuilders can carry on the pleasure of flying . . . purely for sport and for the fun of it.

Maintaining a record of your aircraft construction . . . materials used . . . covering . . . tubing . . . aluminum . . . steel . . . paints, etc. . . . of the various component parts . . . is very important, especially if the aircraft is sold to another and information is needed to maintain or repair the aircraft in the event of a mishap. From a legal standpoint, it could possibly show sincere responsibility on the part of the builder in the event of a questionable mishap.

The EAA has for a number of years had available to builders an EAA Service Manual which, when completed, is a valuable source of history and information. We encourage you to add this to your EAA Amateur Built Aircraft Log.

A homebuilder's dream world becomes reality at the annual EAA Oshkosh Convention. These scenes of wall-to-wall airplanes — people with airplanes on their mind — smiles on their faces — were taken at the 1990 Convention. Top photo shows the famous archway that leads to the fantastic flight line where every type of aircraft is on display — hundreds of homebuilts of unbelieveable workmanship, innovation and design. The seven-day EAA Oshkosh Convention is the world's largest aviation event.

INTERNAT'L EXPERIMENTAL AIRCRAFT ASSOC.
FLY-IN CONVENTION
OSHKOSH '90

What Is An Amateur-Built Aircraft?

3-1 FEDERAL AVIATION REGULATION (FAR) PART 21.191 DEFINES IT.

Operating an aircraft the major portion of which has been fabricated and assembled by persons who undertook the construction project solely for their own education or recreation.

This category for the certification and operation of amateur-built aircraft includes gliders, free balloons, helicopters and gyroplanes.

3-2 USE OF STANDARD COMPONENTS PERMITTED

Does this mean that the sport plane must be constructed in its entirety by the amateur builder? It does not. Commercially produced components and parts which are normally purchased for use in aircraft may be used, including engines and engine accessories, propellors, tires, spring steel landing gear, main and tail rotor blades, rotor hubs, wheel and brake assemblies, forgings, castings, extrusions, and standard aircraft hardware such as pulleys, bellcranks, rod ends, bearings, bolts, rivets, etc. Any kind of materials may be used in the construction of an amateur-built aircraft. Such components and materials should, however, be in good condition and serviceable. Aircraft quality materials are recommended.

3-3 AIRCRAFT QUALITY ESSENTIAL

It is strongly recommended that FAA approved components and established aircraft quality materials (bolts, nuts, pulleys, tubing, extrusions, etc.) be used wherever possible in fabricating parts such as wing spars, critical attachment fittings, and fuselage structural members.

3-4 KEEP IT SAFE

The builder's desire to experiment and to devise new construction methods and techniques should not be done indiscriminately. Heavy reliance on existing standards and established experience factors should be the rule. Any deviation from accepted construction practices and techniques should be verified by stress analysis and/or structural testing of the component or structure.

3-5 CHOOSE A DESIGN CAREFULLY

Deciding on which airplane to build is one of the most important decisions you'll make in your life. It's right up there with picking a mate and buying a home. We advise you to carefully consider the advantages, the drawbacks, and the utility of a wide range of designs before making your decision. Take your time. The majority of amateur aircraft builders have finished and are flying their airplanes . . . and they are very pleased with them. However, there are some who are disappointed with what they have created. Most of the unhappy ones chose their design on the spur of the moment. Also, many aircraft projects have been abandoned long before completion, when the builders came to the realization that they had rushed into the wrong choice. The old adage, "haste makes waste" is very true when it comes to choosing and building an airplane.

While you are evaluating the various designs, remember that building the aircraft will take a great deal of time. You should, if possible, look into the future and try to predict the length of time it will take you to complete your project. Your personal situation may be very different one, two or five years down the line. If you choose an aircraft design based upon your requirements of today . . . you may be disappointed tomorrow. For instance, if a decision is to be made between a one and a two place aircraft, and you are unsure, we recommend selecting the two place aircraft for its future usefulness. You may not have a flying buddy or spouse at this point in time, but you will probably want to share the joy of flight in your own hand-crafted personal sport plane with someone in the future.

Assess your personal building skills in relationship to the complexity of the aircraft projects you are evaluating. The widespread and sometimes superficially written publicity given to homebuilt airplanes has given some newcomers the impression that all designs are easy to build and fly. All of us have become very accustomed to buying mass produced, ready to use or ready to assemble products. We must consciously remind ourselves that building an airplane is a complex project that demands a number of skills.

All too often, an unproven design receives a great deal

Brokaw Bullet, a tandem, all-metal retractable gear speedster was designed and built by Dr. Bergon Brokaw, Leesburg, FL. This high powered, high performance homebuilt is an improved version of the design Dr. Brokaw first introduced in the fall of 1972.

The Hatz Biplane is a simple, straight-forward design incorporating traditional welded steel tube fuselage, wood wing and fabric covering. Most Hatz Biplanes have flat four engines installed, but Ken Williams, an expert on "little round engines", mounted a rare LeBlond 90 Model 5F, a 5-cylinder radial—plus a beautiful, shiny Hamilton Standard ground adjustable steel propeller! With this engine and propeller combination, the Hatz takes on the appearance of classic 30s antique — the type of aircraft that inspired many of today's homebuilders.

of publicity. The plans may not be fully developed . . . or perhaps the prototype hasn't even flown. Choose your design carefully and remember, "caveat emptor" — let the buyer beware.

In choosing a design, the builder must also consider his or her flying skills. Many experimental/amateur-built aircraft have very sensitive controls. Some have very high airspeeds or are capable of unlimited aerobatics. It is unwise and unsafe to build an airplane that demands skills that are beyond your capabilities.

The best reference for an airplane that you'd like to build is the builder who has built and flown one. He can tell you if there are any problems with the plans or the kit and he can give you excellent advice as to the flying and handling characteristics of the airplane. In today's legal environment, it is also a good idea to check with your aviation insurance company through EAA Headquarters to see if it is feasible to insure the aircraft you are planning to build. You may also want to ask EAA Headquarters for information on the safety record of the design. Additionally, EAA Headquarters has lists of articles on amateur-built aircraft that have appeared in its publications. This is an excellent reference source. We recommend that you contact EAA Headquarters for additional information before making your final choice.

In the final analysis, each amateur-built aircraft design has to be given separate and individual evaluation. Some designs have been built in considerable numbers and their flying qualities are well known, their structural soundness has been proven and any bugs in the airplane or the plans have been discovered and taken care of. Other designs are so new that there is not really much known about them. Of course, the aircraft that you chose should be tailored to your needs, your skills and your abilities. The staff at EAA Headquarters stands ready to provide you with information on most designs that will be helpful. Picking the right homebuilt design is a difficult decision. But, the right decision will result in years of satisfying, safe and happy building and flying.

Typical Events In The Building Of A Sportplane

4-1 DECIDE WHAT AIRCRAFT TO BUILD

4-2 PURCHASE PLANS (OR DESIGN YOUR OWN).

Resolve that you will follow the plans faithfully and that you will not make major modification unless qualified. If you can't conscientiously do this . . . design your own! One of the principle causes of safety problems with amateur built aircraft is that modifications to existing plans without the designers consent can quite frequently turn out to be understrength.

4-3 JOIN THE EAA . . . CONTACT YOUR NEAREST EAA CHAPTER AND EAA TECHNICAL COUNSELOR.

4-4 REVIEW YOUR PLANS WITH AN EAA TECHNICAL COUNSELOR OR SOME OTHER SUCCESSFUL BUILDER.

The Technical Counselor or builder will probably know of material sources, availability or parts and jigs, etc.

Others who are building aircraft in your area should be helpful to you, too. Contract EAA Headquarters for a list of SPORT AVIATION articles on the aircraft.

4-5 PURCHASE YOUR CONSTRUCTION MATERIALS AND REFERENCES.

The advertisements in Sport Aviation Magazine are good sources for materials. Custom aircraft log books, EAA Service Manual for your aircraft and "How to Do" construction manuals related to the type of project are available through EAA Air Museum Foundation.

4-6 CONSTRUCT WORK TABLE, SHARPEN TOOLS, SET UP WORKSHOP SPACE

Most builders have to establish a budget for their airplane project and it is quite easy to overbuy on tools. If you purchase tools carefully, and only buy those that you need as you need them, you can stretch your airplane building budget a longer way. However, good lights for the workshop are an absolute essential, as is the ability to heat the workshop for certain projects, such as gluing, fiberglassing and painting.

4-7 ESTABLISH CONTACT WITH YOUR LOCAL FAA

Manufacturing Inspection District Office (MIDO), General Aviation District Office (GADO), or Flight Standards District Office (FSDO) for FAA GUIDANCE AND REQUIREMENTS. Check your EAA Chapter for proper contact in your locale.

4-8 CONSTRUCTION PERIOD USUALLY SPANS TWO TO FIVE YEARS OF EDUCATION AND RECREATION.

4-9 INSPECTIONS OF THE AIRCRAFT

You are responsible for the progressive inspections on the airplane, and you should set up a schedule for this. Since 1979 there has been one final inspection prior to flight. EAA Headquarters recommends a 3-ring binder in which you keep a running daily diary of the aircraft constrution such as: "January 20, 1991, Spent 2 hours on ribs" and also keeping photographs showing you, with your airplane, in your shop. Also keep your bills in that notebok. The purpose of this log book is to prove to the FAA that you are the "major portion" builder of the airplane. Since there is only one final inspection prior to flight this is highly important. A pre cover inspection of each major component should be done by yourself and you should sign the log books. EAA Technical counselors or knowledgeable A&P's may look at the work from time to time and you should make an entry when they stop by such as: "John Smith, Technical Counselor visited Tom Jones' project (such and such a date) signed Tom Jones." You will realize that most A&P's and all Technical Counselors will be reluctant to sign your log books. The final inspection for flight approval is made by the FAA after you request the inspector to come out. The FAA may have to call on a commercial engineer, known as a designated airworthiness representative, to do the final inspection. The FAA does not charge for the final inspection, whereas the DAR is a commercial operator, and he will charge for this.

Offered as a kit with premolded composite parts, the Cirrus VK-30 is an advanced design, high performance, four-place pusher. The prototype was powered with a Lycoming IO-540-G1A5. Aircraft can be fitted with a turbine or V-8 auto engine conversion.

It is best to have an EAA Technical Counselor look your work over from time to time, and in particular, prior to calling the FAA for the pre-flight inspection.

4-10 COVERING AND PAINTING OF THE AIRCRAFT.

(See EAA manuals on covering).

4-11 OBTAINING THE AIRCRAFT REGISTRATION NUMBERS AND AFFIXING THEM TO THE AIRCRAFT.

See Section 9.

4-12 FINAL ASSEMBLY AT THE AIRPORT.

4-13 A CHECK OF THE WEIGHT AND BALANCE OF THE COMPLETED AIRCRAFT.

See Section 14. You might like to contact EAA Headquarters for forms to use for your weight and balance check.

4-14 CONDUCTING THE TAXI TESTS.

Taxiing your aircraft is permitted prior to the final visit by the FAA. However, many an embarrassed builder has run high speed taxi tests that turned into inadvertent flights. It is recommended that you keep these taxi tests down to a speed below at least ½ of your lift off speed. Use the longest and widest runway available for these tests on days wind conditions permit use of these runways.

4-15 THE REQUEST FOR AN AIRWORTHINESS INSPECTION FOR CERTIFICATION AND THE ISSUANCE OF THE AIRWORTHINESS CERTIFICATE BY THE FAA.

4-16 UNDERTAKING THE INITIAL TEST FLIGHT.

Be sure you have given notice to the fixed base operator and airport manager. You should also contact several friends, so that you'll have at least one car on hand to act as a chase car if necessary. You should seriously consider the availability of emergency vehicles. It is a good idea to have your friends go over the aircraft, (friends, technical counselors, A&Ps) before the test flight, looking for the inevitable missing safety pin, lock nut, etc.

4-17 THE CONTINUED TESTING PERIOD OF 25 HOURS OR MORE IN THE ASSIGNED TEST AREA IF TYPE CERTIFICATED ENGINE AND PROPELLER ARE USED. NON-TYPE CERTIFICATE ENGINE OR PROPELLER REQUIRES 40 HOUR TESTING.

4-18 AT THE END OF THE TESTING PERIOD, MOST OPERATING LIMITATIONS WILL ALLOW YOU TO GO ON, ON YOUR HONOR, TO YOUR SECOND PHASE OF OPERATING LIMITATIONS. YOU CAN FLY OUT OF THE TEST AREA ONCE OPERATING LIMITATIONS PERMIT.

4-19 FLY ANYWHERE, ANYTIME, IN ACCORDANCE WITH YOUR OPERATING LIMITATIONS.

4-20 ANNUAL CONDITION INSPECTIONS BECOME FAMILIAR WITH THE YEARS!

Steel tube, wood and fabric biplanes are favorites of homebuilders. The Christian Eagle shown here is available as a kit plane with many components pre-built. The builder must, however, be responsible for 51% of the total construction. Larry Denning, Kansas City, MO won a major trophy at the 1984 EAA Oshkosh convention.

An Introduction To Federal Requirements

5-1 APPLICABILITY

The regulatory and rule making authority of the federal government as it applies to aviation matters is vested in the Federal Aviation Administration (FAA). The FAA regulates both the individual and the machine through its Federal Aviation Regulations (FAR's).

The Federal flight rules and regulations that apply to the person (pilot) are the same whether one flys a manufactured aircraft or a homebuilt certificated in the Experimental category. As that sort of information is easily available, it will not be discussed in detail herein. (FAR Part 61, Certification: Pilots and Flight Instructors and FAR Part 91 General Operating and Flight Rules).

Other rules and regulations, FAR Part 21, (Certification Procedures for Products and Parts) and FAR Part 23 (Airworthiness Standards: Normal, Utility, and Aerobatic Category Airplanes), for example spell out in precise detail the standards by which an aircraft will be manufactured and certificated for sale and use by the public.

No such federal standards exist nor are any imposed on amateur-built aircraft. Nevertheless, no aircraft may fly in the U.S.A. until it has been certificated by the federal government (FAA). This naturally includes aircraft built by the individual at home.

5-2 TYPES OF EXPERIMENTAL CERTIFICATES

To permit a variety of special purpose aircraft to be flown, the FAA identifies, under FAR Part 21.191, the types of certificates which may be issued under the catch-all category of Experimental Certificates. These include:

1. Research and development
2. Showing compliance with regulations
3. Crew training
4. Exhibition
5. Air racing
6. Market surveys
7. **Operating amateur-built aircraft** (This is the one applicable to our purposes) **"Operating an aircraft the major portion of which has been fabricated and assembled by persons who undertook the construction project solely for their own education or recreation."** (FAR Part 21.191) As of this writing, the Amateur-Built Aircraft continue to be certificated under the EXPERIMENTAL label rather than a more appropriate classification such as "Custom-Built Aircraft."

In addition to this minimum guidance afforded by Part 21, there are several FAA Advisory Circulars which outline specific policies governing the amateur builder. Applicable requirements and a listing of the necessary Regulations and Advisory Circulars are listed at the end of this chapter.

Nowhere in the regulations, however, are there detailed airworthiness standards prescribed for the amateur-built aircraft as there are for Normal, Utility, and Aerobatic Category Certificated Aircraft (Part 23).

The FAA Inspector treats each individual application and aircraft separately and the various policies serve as guides only, to give the amateur designer and builder an idea of what the FAA requires of an Amateur-Built aircraft. If the builder complies wherever possible with what few requirements do exist, and observes the traditional standards for aircraft construction, little difficulty should be had in obtaining approval of his amateur-built aircraft. By traditional standards we mean the aircraft should not have any apparent unsatisfactory features of design and construction. In other words, past experience and practices in aircraft design and construction are used as a guide by the FAA Inspectors when looking over a homebuilt aircraft.

Should the builder use radical or unique, untried, and untested materials and construction techniques the burden of proof for the integrity of the structure lies with the homebuilder and the Inspector will have to be convinced. However, in return for these traditional "policies" being followed, and after the inspections and flight test requirements are fulfilled, many of the restrictions that normally apply to other experimental aircraft categories for the duration of their certificates, are removed for the amateur-built aircraft.

5-3 DURATION OF CERTIFICATES

Experimental certificates are issued to all civil aircraft at the beginning of their aeronautical lives upon the owner's request. These experimental certificates are usually issued for an unlimited duration.

5-4 OPERATIONS LIMITATIONS

The FAA can re-inspect an experimental aircraft from time to time and amend, modify, suspend or revoke the certificate. Certain restrictions appropriate to the operation of the aircraft can be made by the Administrator or his representative.

Restrictions on experimental aircraft are usually of a general nature applied primarily to protect the people on the ground rather than the aircraft and its occupants.

All experimental aircraft are restricted against carrying persons or property **for hire or compensation** and this restriction is never removed.

These aircraft are restricted initially to flights within certain approved flight-tested areas. Flights over densely populated areas located within this zone are generally further prohibited. Takeoffs and landings may be excluded in this restriction at the discretion of the FAA inspector.

The greatest difference between the Experimental Amateur-Built and the other experimental sub-categories is that the restrictions concerning passenger carrying and limited flight areas may be removed after a period of flight testing.

The decision as to whether or not any flight restrictions will be removed, however, is entirely up to the FAA Inspector.

Night and IFR flight can be permitted by the FAA inspector, providing the amateur built aircraft is equipped with the same lights and equipment specified for type certificated aircraft. This includes yellow-tagged instruments, TSOd equipment or the operational equivalent, etc. The builder must ask for an operation limitation to allow night and instrument flight.

5-5 REFERENCE SOURCES

The following Federal Aviation Regulations and Advisory Circulars are pertinent to the construction and operation of amateur-built aircraft and may be obtained from any of the GPO (Government Printing Office) book stores located throughout the United States or directly from:

The Superintendent of Documents
U.S. Government Printing Office
Washington, D. C. 20402

5-6 FEDERAL AVIATION REGULATION (FAR's)

1. **CFR Part 14**
   Code of Federal Regulation, Part 14.
   (Aeronautics & Space)
   Parts 1 to 59

2. **CFR Part 14**
   Code of Federal Regulations, Part 14
   (Aeronautics & Space)
   Parts 60 to 139

NOTE: Code of Federal Regulations & FAR are one and the same. The books which are updated annually "Code of Federal Regulations" are about 1/10th the cost of the FARs but do not include proposed changes. However the CFR Part 14 is a much more convenient size, and is the best buy for the amateur builder. A new addition is generally published in March.

5-7 FAA ADVISORY CIRCULARS. (Should read as follows:)

*AC-20-27-D - Certification & Operation of Amateur-Built Aircraft (6-22-90)
AC-65-23 - Certification of Repairmen (Experimental Aircraft Builders)
*AC-20-35C - Tie-Down Sense (7-12-83)
*AC-20-37C - Aircraft Metal Propeller Blade Failure (7-15-80)
*AC-21-12 - Application for U.S. Airworthiness Certificate FAA Form 8130-6
*AC-45-2 - Identification and Registration Marking (7-7-72)
*AC-20-44 - Glass Fiber Fabric for Aircraft Covering (9-3-65)

*These items may be obtained at no charge from:
Department of Transportation, FAA,
U.S. Department of Transportation
Subsequent Distribution Section, M-494.3
Washington, D.C. 20590

AC-43.13-1A - Acceptable Methods, Techniques and Practices - Aircraft Inspection and Repair (4-17-72). (AWS-340)

Contains methods, techniques, and practices acceptable to the Administrator for inspection and repair to civil aircraft. Published in 1973. (Consolidated reprint incorporates Changes 1 and 2.) ($8.00 - Supt. Docs.) SN 050-007-00461-3.

AC-43.13-2A - Acceptable Methods, Techniques and Practices - Aircraft Alterations (6-9-77). (AWS-340).

Contains methods, techniques, and practices acceptable to the Administrator for use in altering civil aircraft. ($5.50 Supt. Docs.) SN 050-007-00625-0. (Includes Chg. 1).

NOTE: These two circulars (AC-43.13) replace old Cam 18 which is no longer available from federal sources.

"FAA Publications", Free, DOT, Distribution Unit TAD 484.3, Washington, D.C. 20590

AIM: see "FAA Publications" free brochure for this and other publications.

FEDERAL AIR REGULATIONS
Applicable To Amateur-Built Aircraft

PART 21	**Certification**
21.93	Classification of minor and major changes to experimental aircraft.
21.171	Applicability
21.173	Eligibility
21.175	Special Airworthiness Certificate
21.177	Amendments of special airworthiness
21.179	Transfer of certificates
21.181	Duration of special airworthiness certificates
21.182	Aircraft ID required
21.191	Experimental certificates, amateur-built
21.193	Experimental general (letter required)
21.275	Limitations and conditions on experimental certificates
21.475	Limitations and conditions on experimental certificates
PART 39	**A.D. Notes**
	These notes can be applied to experimental aircraft - are not **usually.**
PART 43	**Maintenance, Preventive Maintenance, Etc.**
43.13B	Part 43 does not apply to experimental aircraft, it is highly important that people know this.
PART 45	**Identification and Registration Marking**
45.1	Identification of aircraft and engines
45.11	Identification of aircraft and engines and also plate required
45.13	Identification Data & Plates
45.21	General
45.22	Exhibition, Antique & Other Special Rules
45.23	Experimental Number Size, Display of Marks
45.25	Location of Marks
45.27	Location of Marks, non-fixed wing aircraft
45.29	Size of marks, small N numbers ok'd
PART 47	**Aircraft Registration**
47.1	Applicability
47.3	Required Registration
47.5	Applicants
47.7	Who can register an aircraft
47.11	Evidence of Ownership
47.15	What ID numbers can be used
47.17	Fees
47.19	Registry Address
47.31	Certificate Application
47.33	Certificate Registration
47.39	Effective Date of Registration
47.41	Duration and Return of Certificate
47.45	Change of Address
PART 49	**Recording of Aircraft Titles & Security Documents**
49.11	Addresses
49.15	Fees
PART 61	**Certification of Pilots & Flight Instructors**
61.31 (h) (3)	Rating not required for Experimental aircraft unless operating limits specifically require it.

PART 65	**Certification of Airmen Other Than Crew Members (Repairman's Certificates)**
65.1	Applicability
65.11	Application and Issue
65.101	Repairman, Eligibility Requirements
65.103	Repairman's Certificate, Privileges & Limitations
65.104	Repairman's Certificate, Experimental Aircraft Builder
65.105	Display of Repairman's Certificate
PART 91	**General Operation and Flight Rules**
91.7	Civil Aircraft Airworthiness
91.9	Operating limits and marking requirements
91.101	Applicability
91.171	VOR equipment check for IFR flight
91.203	Certifications required
91.205	Instruments and equipment required
91.207	Emergency Locator Transmitters
91.215	Mode C to Mode S 7-1-92
91.319	Operating limits for experimental aircraft
91.413	ATC transponder tests and inspections required every two years

§91.42 Aircraft having experimental certificates: operating limitations.

(a) No person may operate an aircraft that has an experimental certificate:

(1) For other than the purpose for which the certificate was issued: or

(2) Carrying persons or property for compensation or hire.

(b) No person may operate an aircraft that has an experimental certificate outside of an area assigned by the Administrator until it is shown that:

(1) The aircraft is controllable throughout its normal range of speeds and throughout all the maneuvers to be executed; and

(2) The aircraft has no hazardous operating characteristics or design features.

(c) Unless otherwise authorized by the Administrator in special operating limitations, no person may operate an aircraft that has an experimental certificate over a densely populated area or in a congested airway. The Administrator may issue special operating limitations for particular aircraft to permit takeoffs and landings to be conducted over a densely populated area or in a congested airway, in accordance with terms and conditions specified in the authorization in the interest of safety in air commerce.

(d) Each person operating an aircraft that has an experimental certificate shall:

(1) Advise each person carried of the experimental nature of the aircraft;

(2) Operate under VFR, day only, unless otherwise specifically authorized by the Administrator: and

(3) Notify the control tower of the experimental nature of the aircraft when operating the aircraft into or out of airports with operating control towers.

(e) The Administrator may prescribe additional limitations that he considers necessary, including limitations on the persons that may be carried in the aircraft.

(49 U.S.C. 1423)
[Amdt. 91-53, 33 FR 6859, May 7, 1968, as amended by Amdt. 91-109, 38 FR 1176, Jan. 10, 1973]

On the foreign country rules, contact the following for rules and regulations in these countries:

Great Britain
Popular Flying Assoc.
Terminal Building, Shoreham Airport
Shorham By Sea, Sussex
England BN4 5FF

France
The RSA
Reseau Du Sport de L'Air
Siege 39, Rue Sauffroy
75017 Paris France

Australia
The Sport Aircraft Assoc. of Australia
Federal Council
President Peter Crooks
A. Gilbirch Clinic
195 Gilbert Rd.
West Preston 3072
Australia

New Zealand
The New Zealand Amateur Aircraft
Constructors Assoc.
P.O. Box 6417
Te Aro Wellington, New Zealand

Homebuilt aircraft take to the air with the simplest lightweight ultralight engines to the most sophisticated engine installations such as this 350 hp Continental TSIO-550A installed on a Lancair IV. With this powerplant the 4-place Lancair at 75% power at 25,000 ft. has a cruise speed of 331 mph.

FAA Inspections

6-1 GET ACQUAINTED WITH THE FAA

An important phase of the airworthiness certification procedure includes inspection of the aircraft by an authorized FAA representative. Either your local FAA representative from the Engineering Manufacturing District Office (EMDO) or a private party known as a Designated Airworthiness Representative (DAR) may do the one final inspection and determine that an amateur-built aircraft is in condition for safe operation. The FAA inspection is free, the DAR is not. If the FAA can't do the inspection, the DAR will charge for travel, his time and insurance — $100-$300 is usual.

An excellent starting point for you if you are a first-time builder is not the hasty ordering of materials but rather an early contact with EAA and GADO, FSDO, or MIDO of the FAA nearest your home. The purpose of such a contact would be to discuss with the FAA representative, your proposed homebuilt project and to generally familiarize yourself with the procedures established by the FAA for homebuilt projects. At this time you can establish a tentative plan for inspection of the aircraft upon completion. A FAA representative may also have other information you will find useful.

You can accept this as an absolute fact: the typical FAA inspector is interested in your project and wants to help you do a good job.

6-2 AIRCRAFT LOGBOOK

Everyone building an airplane must maintain a construction log of the work he does on his airplane. Most builders have found the usual small aircraft log too small a book to use for this purpose. You can use a notebook of conventional size and keep a daily diary of work done on the aircraft. It is also suggested that you keep photographs of work done on the airplane and, in particular, showing you working on the aircraft in various stages of construction in this book. Your bills of sale could also be maintained in this book in an organized manner. Advisory Circular 20-27C available from the FAA or EAA describes the procedure used so that your log book will be a verification of your having completed at least 51% of the aircraft yourself. Upon completion of the aircraft, it is suggested that you obtain a normal aircraft log, an engine log and a propeller log and continue your logs from construction into these. By this means, you will not need to carry a large notebook in the aircraft after it is inspected and flown. The FAR's do require that you keep a log book on the aircraft, on the engine and on the propeller. This can either be one log book with a recorded history of all three or three different smaller log books on each item.

6-3 TYPICAL LOG BOOK ENTRIES

In your construction log book, you can make entries to reflect certain stages of progress and the inspections you, the builder, have made on the aircraft before covering or closing any major component. The FAA EMDO or GADO inspector will also make his entry in the log book upon making his one inspection to the aircraft prior to flight.

6-4 HOW MANY INSPECTIONS?

The builder should establish an inspection schedule and inspect his own aircraft from time to time during its construction. The EAA Technical Counselor does not make actual "inspections", but will visit the project from time to time. The Technical Counselor's position is that of a friend visiting a friend, making an informal visit to the project to help the builder look for anything he may not have noticed. It is always good to have an extra set of eyes looking over your work. Most chapters have Technical Counselors that are experienced A&P mechanics or aircraft builders that can provide this service. However, the builder is totally responsible for the airworthiness of his aircraft and the Technical Counselor does not sign log books or confirm the airworthiness of the aircraft. This responsibility is totally up to the builder, and he is expected to complete a totally airworthy aircraft prior to the final inspection by the FAA.

Those builders who are already building their new plane and have not yet conversed with the FAA about the required inspection, should get in touch with the inspector as soon as possible. Figure 6-1 shows a sample letter requesting an inspection.

6-6 KEEP ALL RECEIPTS TO ESTABLISH OWNERSHIP

A number of log book pages may be reserved for recording materials and sources of supply of the major construction items such as: spar and capstrip material; sheet metal; plywood; glue; tubing; wheels; brakes; etc. Receipts of all such purchases should be retained until the list has been presented to the FAA Inspector for confirmation, if requested.

6-7 THE EAA SERVICE & MAINTENANCE MANUAL

A preferred arrangement, other than making a lot of crowded entries in the Aircraft Log Book, would be for the builder to acquire an EAA Service & Maintenance Manual and to make all of the applicable entries in it. The service manual, specifically designed for the homebuilder, is intended to serve as an erection and maintenance manual and an operations manual for that aircraft. A service manual containing the information that the builder considers essential for proper maintenance should be made available to the FAA inspector at the time of certification of the airplane.

The builder should consider at least the following in developing the essential information for the service manual:

Aircraft specifications and weight and balance.

Description of systems such as electrical, hydraulic, fuel controls, etc.

Lubrication instructions setting forth the frequency, and

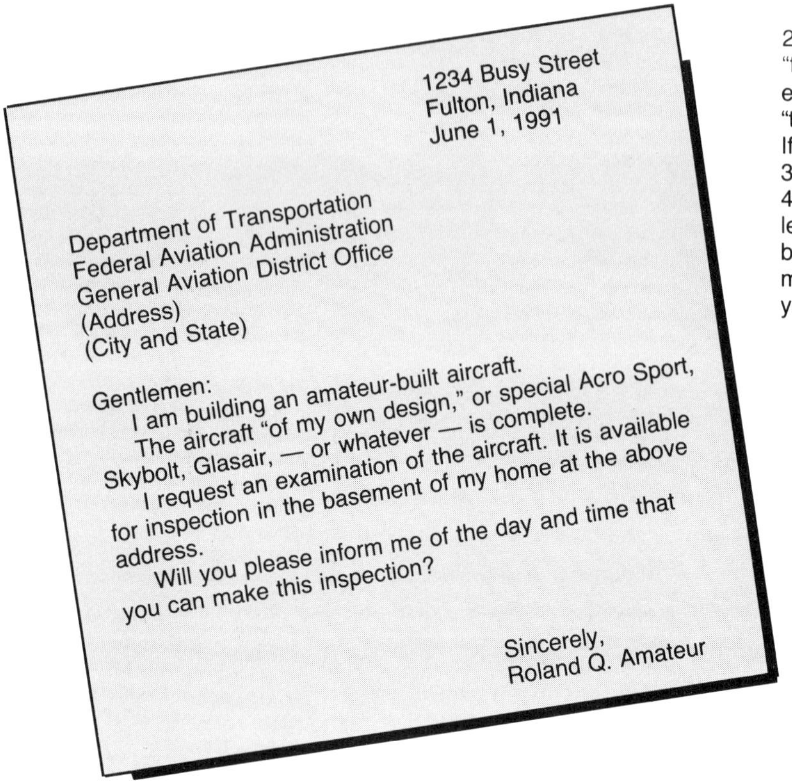

1234 Busy Street
Fulton, Indiana
June 1, 1991

Department of Transportation
Federal Aviation Administration
General Aviation District Office
(Address)
(City and State)

Gentlemen:

I am building an amateur-built aircraft.

The aircraft "of my own design," or special Acro Sport, Skybolt, Glasair, — or whatever — is complete.

I request an examination of the aircraft. It is available for inspection in the basement of my home at the above address.

Will you please inform me of the day and time that you can make this inspection?

Sincerely,
Roland Q. Amateur

In your letter include all the information required by FAR 21.193. You should describe the fact that the aircraft was built "for my education and recreation", and that the "nature of the experiment is flying amateur built aircraft." You should state "the aircraft type", if it is readily identifiable like a Pitts Special. If not you should state that "I am enclosing a photograph or 3-view to identify the aircraft." You should also request the 40 hour test period if non type certificated engine and propeller or 25 hours if type certificated engine and propeller combined. You should also request a test area, usually within 25 miles of your home base and describe the area over which you will be flying.

FIGURE 6-1

SAMPLE LETTER:
Requesting an Inspection at End of Construction

FAA inspections are normally conducted by one inspector. It was a special occasion that brought this group of inspectors together for a thorough FAA going-over of the EAA's "Spirit of St. Louis" replica and its first flight.

the lubricants, and fluids, which are to be used in the various systems.

Pressures and electrical loads applicable to the various systems.

Tolerances and adjustments necessary for proper functioning of the airplane.

Methods of leveling, raising, and towing.

Methods of balancing control surfaces.

Identification of primary and secondary structures.

Frequency and extent of inspections necessary for proper maintenance of the airplane.

It can be readily apparent that a properly completed EAA Service Manual would become very valuable to the owner and in particular future owners of the aircraft and help assure its proper future maintenance and operation. Service Manuals are available from EAA.

In addition to the log books and the EAA Service Manual which now record material sources, pre-cover inspections, and similar information, there are two items that must be accomplished, namely, the equipment list and the weight and balance data form. Most aircraft logs conveniently include data or memorandum pages in the back which are made to order for equipment lists. A single typed page including all of the weight and balance data, stapled in the aircraft log will provide all of the required aircraft's papers in one neat package.

Both the equipment list and the weight and balance data pages should be dated and identified by listing the make, model, serial number, and registration number of the aircraft.

6-8 EQUIPMENT LIST

The equipment list is composed of a list of all the items of equipment which were installed in the plane at the time of weighing, including make and model of removable or special items such as propeller, engine, starter, generator, battery, instruments, wheels, tires, radio, etc. When future changes in equipment are made, the equipment list must be changed to reflect the addition or deletion of equipment and the date of the change.

6-9 WEIGHT AND BALANCE DATA

The weight and balance data record is a sheet upon which the following points are located and the computations of weight are shown: Location of datum: location of three weighing points; actual empty weight of aircraft; fuel and oil capacity and location; computed gross weight of aircraft; baggage compartment capacity and location; number of seats and location; actual empty CG and computed loaded CG.

6-10 THE PRE-CERTIFICATION INSPECTION

Your local FAA inspector will review all the papers; inspect the complete airplane (inspection plates and cowling removed), see that the registration certificate is displayed in the plane, check for proper instrument markings and cockpit placards, (see section 8).

If you have presented complete and orderly records and a "clean" aircraft, the inspector will present you with an Airworthiness Certificate and your Operating Limitations Form. He will then make an entry in your log book authorizing the commencement of flight testing followed by his signature. The builder is encouraged to taxi test prior to FAA inspection - but flight is prohibited until the special airworthiness is signed by the FAA inspector.

The inspector will issue an unlimited airworthiness certificate upon completion of the pre-certification inspection, if everything goes well. He will provide you with two sets of operating limitations, or operating limitations listed as Phase I and Phase II. The Phase I operating limitations cover you during the initial testing period. When this period is completed, you are on your honor to commence operations under the Phase II operating limitations. This is done to eliminate paper work back and forth between you and the FAA at the end of the testing period, and is a sign of the confidence placed in the aircraft builders by the FAA, a trust that we must keep for the sake of other aircraft builders.

FAR 21.191 Experimental certificates.

Experimental certificates are issued for the following purposes:

(a) *Research and development.*

(b) *Showing compliance with regulations.*

(c) *Crew training.*

(d) *Exhibition.*

(e) *Air racing.*

(f) *Market surveys.*

(g) *Operating amateur-built aircraft.* Operating an aircraft the major portion of which has been fabricated and assembled by persons who undertook the construction project solely for their own education or recreation.

FAR 21.193 Experimental certificates: general.

An applicant for an experimental certificate must submit the following information:

(a) A statement, in a form and manner prescribed by the Administrator setting forth the purpose for which the aircraft is to be used.

(b) Enough data (such as photographs) to identify the aircraft.

(c) Upon inspection of the aircraft, any pertinent information found necessary by the Administrator to safeguard the general public.

(d) In the case of an aircraft to be used for experimental purposes—

(1) The purpose of the experiment;

(2) The estimated time or number of flights required for the experiment;

(3) The areas over which the experiment will be conducted; and

(4) Except for aircraft converted from a previously certificated type without appreciable change in the external configuration, three-view drawings or three-view dimensioned photographs of the aircraft.

FAR 21.93 Classification of changes in type design.

Changes in type design are classified as minor and major. A "minor change" is one that has no appreciable effect on the weight, balance, structural strength, reliability, operational characteristics, or other characteristics affecting the airworthiness of the product. All other changes are "major changes".

Part 43 — MAINTENANCE, PREVENTIVE MAINTENANCE, REBUILDING, AND ALTERATION

(b) This part does not apply to any aircraft for which an experimental aircraft airworthiness certificate has been issued, unless a different kind of airworthiness certificate had previously been issued for that aircraft.

NOTE: Although this is a list of manufacturing inspection offices, airworthiness certification may be accomplished by the applicants nearest general aviation district office (GADO) or flight standards district office (FSDO).

NEW ENGLAND REGION

Manufacturing Inspection District Office
NE-MIDO-41
Bradley International Airport
Combs-Gates Building
Windsor Locks, Connecticut 06096
COMM: (203) 627-0225/0226

EASTERN REGION

Manufacturing Inspection District Office
NE-MIDO-44
Federal Building
P.O. Box 640, 228 Walnut Street
Harrisburg, Pennsylvania 17108
COMM: (717) 782-4425

Manufacturing Inspection District Office
NE-MIDO-45
150 Riser Road
Teterboro Airport
Teterboro, New Jersey 07608
COMM: (201) 288-4340

Manufacturing Inspection District Office
NE-MIDO-46
DOT Administration Building, 2nd Floor
Republic Airport, Route 110
East Farmingdale, New York 11735
COMM: (516) 694-8420

SOUTHERN REGION

Atlanta MIDO
1075 Innerloop Road
College Park, Georgia 30337
COMM: (404) 763-7621

Savannah MISO
Rt. 1, Box 288D
Savannah, Georgia 31408
COMM: (912) 964-9017

Mobile MISO
Building 28, Brookley Field
P.O. Box 5196, Fulton Rd. Station
Mobile, Alabama 36615
COMM: (205) 690-2253

Nashville MISO
322 Knapp Boulevard, Room 101
Nashville Metro Airport
Nashville, Tennessee 37217
COMM: (615) 251-5661 & (615) 251-5196

Listed below are three offices that generally do amateur-built aircraft inspections. Contact your local office.

1. GADO (General Aviation District Office)
2. FSDO (Flight Information Service)
3. MIDO (Manufacturers Inspection District Office)

If you should have further questions that the above three offices are unable to help you with, contact your regional office, addresses and telephone numbers listed below.

ALASKAN REGION
Federal Aviation Administration
701 C Street, Box 14
Anchorage, AK 99513
907-271-5645

CENTRAL REGION
Federal Aviation Administration
601 East 12th St.
Federal Building
Kansas City, MO 64106
816-426-5626

EASTERN REGION
Federal Aviation Administration
JFK International Airport
Fitzgerald Federal Building
Jamaica, NY 11430
718-917-1005

GREAT LAKES REGION
Federal Aviation Administration
O'Hare Lake Office Center
2300 East Devon Avenue
Des Plaines, IL 60018
312-694-7000

NEW ENGLAND REGION
Federal Aviation Administration
12 New England Executive Park
Burlington, MA 01803
617-273-7244

NORTHWEST MOUNTAIN REGION
Federal Aviation Administration
17900 Pacific Highway South
C-68966
Seattle, WA 98168
206-431-2001

SOUTHERN REGION
Federal Aviation Administration
3400 Norman Berry Drive
East Point, GA
Mail Address: P.O. Box 20636
Atlanta, GA 30320
404-763-7222

SOUTHWEST REGION
Federal Aviation Administration
4400 Blue Mound Road
Mail Address: P.O. Box 1689
Fort Worth, TX 76101
817-734-25001

WESTERN-PACIFIC REGION
Federal Aviation Administration
15000 Aviation Blvd.
Hawthorne, CA
Mail Address: P.O. Box 92007
Worldway Postal Center
Los Angeles, CA 90009
213-297-1427

EUROPEAN OFFICE HEADQUARTERS
(Europe, Africa & Middle East Office)
Federal Aviation Administration
15, Rue de la Loi (3rd Floor)
B-1040 Brussels, Belgium
Mail Address: c/o American Embassy
APO, New York 09667
(32-2) 513.38.30 ext. 2701

Miami MIDO
Room 200, FAA/NWS Building
Miami International Airport
P.O. Box 592373
Miami, Florida 33159
COMM: (305) 526-2600

Vero Beach Manufacturing Inspection Satellite Office
Vero Beach Airport
P.O. Box 2782
Vero Beach, Florida 32960
COMM: (305) 562-0110

Lakeland Manufacturing Inspection Satellite Office
Terminal Bldg., Office #119
Lakeland Municipal Airport
3450 Dranefield Rd., P.O. Box 6133
Lakeland, Florida 33803
COMM: (813) 644-1414

GREAT LAKES REGION

Minneapolis Manufacturing Inspection District Office
6201 34th Avenue South
Minneapolis, Minnesota 55450
COMM: (612) 725-3373

Chicago Manufacturing Inspection Satellite Office
2300 East Devon Avenue, Room 232
Des Plaines, Illinois 60018
COMM: (312) 694-7190

Cleveland Manufacturing Inspection District Office
Federal Facilities Building, Room 127
Cleveland Hopkins International Airport
Cleveland, Ohio 44135
COMM: (216) 522-4140

Detroit Manufacturing Inspection Satellite Office
Willow Run Airport - East Side
8800 Beck Road
Belleville, MI 48111
COMM: (313) 482-1601

Vandalia Manufacturing Inspection District Office
3800 Wright Drive
Vandalia, Ohio 45377
COMM: (513) 898-3991

Indianapolis Manufacturing Inspection Satellite Office
6801 Pierson Drive
Indianapolis, Indiana 46241
COMM: (317) 247-2425

CENTRAL REGION

Wichita Manufacturing Inspection District Office
1801 Airport Road, Room 101
Mid-Continent Airport
Wichita, Kansas 67209
COMM: (316) 946-4450

Kansas City Manufacturing Inspection District Office
Downtown Airport, Room 227
250 Richards Road
Kansas City, Missouri 64116
COMM: (816) 374-2492

SOUTHWEST REGION

SW-MIDO-41
Wiley Post Airport
Terminal Bldg., 2nd Floor
Bethany, Oklahoma 73008
COMM: (405) 787-3842

SW-MIDO-42
4400 Blue Mound Road
P.O. Box 1689
Fort Worth, Texas 76101
(817) 877-2559

ASW-MIDO-43
Room 100, Chaparral Aviation Bldg.
1115 Paul Wilkins Road
San Antonio, Texas 78216
COMM: (512) 824-7560/69

NORTHWEST MOUNTAIN REGION

Manufacturing Inspection Branch - ANM-180S
7300 Perimeter Road South
Seattle, Washington 98108
COMM: (206) 431-2838

Denver Aircraft Certification Office - ANM-170D
10455 East 25th Avenue
Aurora, Colorado 80010
COMM: (303) 340-5575

ALASKAN REGION

Anchorage Aircraft Certification Office - ANM-170A
Federal Building
P.O. Box 14, 701 C Street
Anchorage, AK 99513
COMM: (907) 271-5927

WESTERN PACIFIC REGION

FAA, Los Angeles Manufacturing Inspection District Office - NM-MIDO-46
5885 West Imperial Highway
Los Angeles, CA 90045
COMM: (213) 215-2057

FAA, Phoenix Manufacturing Inspection Satellite Office - MISO-46
15041 North Airport Drive
Scottsdale, AZ 85260
COMM: (602) 241-2112

FAA, Long Beach Manufacturing Inspection District Office - NM-MIDO-47
4340 Donald Douglas Drive
Long Beach, CA 90808
COMM: (213) 421-1759

FAA, Van Nuys Manufacturing Inspection District Office - NM-MIDO-48
7120 Hayvenhurst Avenue
Van Nuys, CA 91406
COMM: (818) 904-6298/99

FAA, San Jose Manufacturing Inspection Satellite Office - NM-MISO-48
1387 Airport Blvd.
San Jose, CA 95110
COMM: (408) 275-7597

CANADIAN
AMATEUR-BUILT REGS SUMMARIZED

From Airworthiness Manual Chapter 549 Airworthiness Standards Amateur-Built Aircraft T51-13/549E

Order from Canadian Government Publishing Centre
Ottawa, Ontario, K1A OS9
Cost is $11.00 plus $3.50 postage plus GST $1.01 = $15.51
Cheque to Receiver General.

Gross Weight: 3,968 pounds
Wing Loading no flaps: 13.3 lb. sq. ft.
Wing loading with flaps: 20.4 lb. sq. ft.
Maximum 4 seats

More than 50% must be fabricated from raw material and assembled by an individual or group of individuals on a non-commercial, non-production basis for educational or recreational purposes. Materials, methods of fabrication and assembly, and workmanship shall be appropriate and should conform to accepted aircraft practices.

NOTE: Amateur-Built Aircraft that exceed the wing loading requirements listed above are classified as High Performance Aircraft. High Performance Aircraft require a cockpit placard and operation requires a High Performance Type Rating for the pilot.

- Inspections required during construction.
- Operates with special Cof A.
- Initial operating restrictions apply.
- A fee for registration marks.
- Minimum instrument requirements.
- Safety belts required.
- Firewall standards.
- Carb. icing prevention needed.
- Various placards needed.
- I.D. plate needed.
- Logs and maintenance records needed.
- Minimum engine rated power requirements.
- Pilot license required.

PRESENT ULTRA-LIGHT REGS SUMMARIZED

From Ultra-light Aeroplane and Hang Glider information manual TP4310E

Order: Transport Canada, AANDHD
Ottawa, Ont. K1A 0N8

Cost is $5.00 plus GST .35¢ = $5.35

Max. launch weight single place: 363 pounds
Max. launch weight two place: 430 pounds
Minimum wing area: 107.6 sq. ft.
Maximum wing loading: 5.14 lb. sq. ft.

- Fee for registration marks.
- I.D. plate required.
- No Cof A or Flight Permit.
- No logs
- No E.L.T.
- Ultra-light pilot license required.

PROPOSED ULTRA-LIGHT REGS

That will come into effect January 1, 1993

Single place gross weight: 628.3 pounds
Two place gross weight: 1058.2 pounds
Maximum stall speed at gross weight: 45 MPH

- Can be used to carry passengers, for rental or training.
- Standards of airworthiness re design and manufacturing.
- Maintenance and continuing airworthiness procedures apply.
- Fee for registration.
- Can be constructed in the Amateur-Built category but same standards as for Amateur-Built Aircraft apply re 50% rule, aircraft quality materials, accepted aircraft practices, etc.
- Can be flown with an Ultra-light Pilot license.

NOTE: These are very minimal summaries. Full details are in the manuals shown. Proposed Ultra-Light regs and policies can be received by writing the EAA Canadian Council, 2348 Garnet St., Regina, Sask. S4T 3A2 or EAA headquarters.

Forms required for Amateur-Built Aircraft
- Application to construct an Amateur-Built Aircraft
- Application for Initial Flight Permit
- Weight Report
- Climb Test Report. Contact nearest Transport Canada Regional Office for information and forms.

REGIONAL OFFICES OF TRANSPORT CANADA AVIATION GROUP (TCAG):

PACIFIC REGION:
Transport Canada
Aviation Group
Box 220
800 Burrard Street
Vancouver, B.C.
V6Z 2J8
604-666-5563

WESTERN REGION:
Transport Canada
Aviation Group
Federal Building
9820 - 107th Street
Edmonton, Alta
T5K 103
403-420-3869

CENTRAL REGION:
Transport Canada
Aviation Group
18-333 Main Street
Box 8550
Winnipeg, Manitoba
R3C OP6
204-983-4341

ONTARIO REGION:
Transport Canada
Aviation Group
4900 Yonge Street
Suite 300
Willowdale, Ontario
M2N 6A5
416-224-3259

QUEBEC REGION:
Transport Canada
Aviation Group
Montreal International Airport
Box 5000
Dorval, Quebec
H4Y 1B9
514-633-3240

ATLANTIC REGION:
Transport Canada
Aviation Group
Box 42
Moncton, New Brunswick
E1C 8K6
506-857-7114

Free Booklet listing CIVIL AVIATION PUBLICATIONS CAN BE OBTAINED BY WRITING:
Transport Canada
AANDHD
Ottawa, Ontario
K1A 0N8
Phone: 613-991-9970
Ask for Catalog Number TP3680E

FUEL FLOW TEST: Mandatory in some Transport Canada Aivation Group Regions. Highly recommended to ensure adequate fuel flow to your engine in nose-high attitudes. Copy available from the EAA Canadian Council, 2348 Garnet Street, Regina, Sask., S4T 3A2 Canada.

EAA Technical Counselors

Volunteers Who Can Provide Assistance From EAA Headquarters Or Its Chapters

We applaud your knowledge of correct design and your talents in flawless workmanship. However, being human, it has been the experience of some builders to find that what they built isn't exactly to aircraft standards, or just doesn't work.

What do you do when you make a mistake in construction or design? How do you know if your ideas and products are right? Is that modification increasing or decreasing structural strength? Are you sure your control linkages are foolproof? Will you trust your life to what you know, or to what you think you know?

If any of these questions apply to you, we suggest that you contact one of these qualified men known as "EAA Technical Counselors".

To be exact, an EAA Technical Counselor is a volunteer member who is dedicated to insuring the safety and dependability of amateur built aircraft. Any one of them will be glad to visit your project, offer constructive advice, suggest changes if necessary, and go over your aircraft with a fine tooth comb. The time spent with an EAA Technical Counselor will pay off in less time correcting mistakes.

EAA Technical Counselors know aircraft. In a survey conducted recently, over 42% have A & P licenses, 83% have built or are building aircraft, 100% have restored or built light aircraft, 24% have completely built more than one aircraft, 94% are pilots with average 1,630 hours of flight time. That's experience in anyone's book.

Since it's inception, the EAA Technical Counselor program has become an important and powerful influence on builders of aircraft. The program has strengthened the already good FAA-EAA relationships and many times EAA Technical Counselors are consulted when accidents do occur. Reaction of the EAA membership has been hearty approval, and the number of EAA Technical Counselor applicants is increasing every month.

The problems discovered by the EAA Technical Counselors have been many. The common mistake found by almost every EAA Technical Counselor in making his rounds has been errors caused by modifications. In some cases, the builder's modifications decreased the safety, strength, and many times, the performance of the craft. Changes and modifications have often resulted in reduced flight performance and reduced tolerable "G" loadings and safety of flight.

A second problem area found by some EAA Technical Counselors has been the use of designs or materials lacking sufficient strength to carry the loads. In past issues of *SPORT AVIATION* this has been pointed out. Wings have come off in flight because of inadequate spars, excessive holes drilled in spars, improper splicing of wooden spars and landing gears have collapsed because of poor mounts. In original designs these mistakes are usually due to lack of knowledge of the designer, but where modifications to existing plans is desired, here the assistance of the EAA Technical Counselor can be invaluable to assure that the strength the designer planned is retained in the design. Rarely has an aircraft been subject to FAA "squawks" when the builder followed the plans to the letter.

If you are building an aircraft, and you have any problems or questions, consult an EAA Technical Counselor! Have him look at your craft prior to final FAA inspection. If you write an EAA Technical Counselor, a courteous gesture would be to enclose a self-addressed, stamped envelope. However, most local chapters have at least one EAA Technical Counselor, EAA headquarters also has complete listings of EAA Technical Counselors in your area, available for the asking.

Build an airworthy ship, friend. Your life depends on it.

Shades of days gone by! Nope, it's not a Cub from the 40s or 50s — this is a brand new amateur built Cuby — plans and many parts readily available to today's homebuilder.

Aircraft Instrument Markings And Cockpit Placards

Your reference is FAR Part 91.31 Civil Aircraft Operating Limitations and Marking Requirements.

8-1 GENERAL

To insure that each person operating an aircraft does so within the operating limitations prescribed for it, the FAA requires that there is available in it a current Flight Manual, appropriate instrument marking and placards, **or any combination thereof.**

The purpose of the flight manual, markings and placards is to detail for the operator of the aircraft, the operational limitations prescribed for the aircraft.

In lieu of a flight manual most amateur builders prefer to mark their instruments and to affix the necessary placards to the instrument panel as the primary means for complying with these requirements.

8-2 MARKINGS AND PLACARDS

The markings and placards necessary for the safe operation and handling of the aircraft should be displayed in a conspicuous place and may not be easily erased, disfigured or obscured. Such placards and markings should include but not necessarily be limited to the following criteria: special emphasis on fuel system markings are very important; such as fuel valves — on-off, fuel octane quantity, unuseable fuel, minimum fuel for take-off, minimum fuel for inverted flight, etc.

8-3 POWERPLANT INSTRUMENT MARKINGS

Each required powerplant instrument should be marked to indicate the maximum and, if applicable, minimum safe operating limit with a red radial line.

Each normal operating range is to be marked with a green arc not extending beyond the maximum and minimum continuous safe operating limits.

Each engine speed range that is restricted because of excessive vibration should be marked with a red arc.

8-4 AIRSPEED INSTRUMENT MARKINGS

The airspeed indicator should be marked with a **radial red line** to establish the never-exceed speed (Vne).

The takeoff and any precautionary range should be marked with a **yellow arc**. The normal range is marked with a **green arc**. The flap actuation range is marked with a **white arc**.

8-5 AIRSPEED PLACARDS

There should be an airspeed placard in clear view of the pilot and as close as practicable to the airspeed indicator listing:

The design maneuvering speed.

The maximum landing gear operating speed (if applicable).

The maximum flap extension operating speed (if applicable).

8-6 LANDING GEARS

If a retractable landing gear is used, an indicator should be marked so that the pilot can, at any time, ascertain that the wheels are secured in their extreme positions.

Each emergency control should be **red** and must be marked as to method of operation and identity.

8-7 CONTROL MARKINGS

Each fuel tank selector should be marked to indicate the position corresponding to each tank and to existing cross feed position.

If safe operation requires the use of any tanks in a specific sequence, that sequence must be identified.

8-8 POWERPLANT FUEL CONTROLS

Each fuel tank selector should be marked to indicate the position corresponding to each tank and to existing cross feed position.

If safe operation requires the use of any tanks in a specific sequence, that sequence must be identified.

8-9 FLIGHT MANEUVER PLACARD

For non-acrobatic category airplanes, there should be a placard in front of and in clear view of the pilot stating: "No acrobatic maneuvers, including spins, approved."

For acrobatic category airplanes, there should be a placard in clear view of the pilot listing the approved acrobatic maneuvers and the recommended entry airspeed for each. If inverted flight maneuvers are not approved, the placard must have a notation to this effect.

8-10 BAGGAGE PLACARD

The maximum baggage load permitted should be displayed in a conspicuous place adjacent to the baggage area.

8-11 PASSENGER WARNING PLACARD

A placard must be affixed to the aircraft so that it is readily seen in the cockpit. It will state: "Passenger Warning — This aircraft is amateur built and does not comply with the Federal Safety Regulations for "Standard Aircraft". This placard is part of a set available from EAA. See Section 10-5.

Left — Even some of the smallest homebuilts have rather complete panels. Builder's choice! Above — Beautiful panel of a Skybolt under construction. Note wires for EGT and CHT switches top right. Note too, neat markings on panel, and well laid out electrical side panel.

A little planning can save a lot of extra work later on. By building an access panel in the front seat bulkhead, as was done on this Acro II biplane, instruments can be reached for easy servicing. Right — Instruments are normally not placed in the center areas of panels of aerobatic aircraft — this space used for posting aerobatic sequence cards. Tube at extreme right of panel is a simple, foolproof sight gauge for fuel.

Aircraft Registration Procedures

Your Reference is FAR Part 47 Aircraft Registration.

9-1 IDENTIFICATION NUMBER

Before a builder or applicant can register his aircraft for the first time, he must obtain an identification number (to be painted on the aircraft) and an Aircraft Registration Application, (AC Form 8050-1). The Aircraft Registration application form can then be completed and submitted to register the aircraft. (Figure 9-5)

Step 1. How To Obtain a Number

How to obtain an identification number: If any identification number is acceptable to you, write to the FAA Aircraft Registry and ask them to assign you a U.S. identification number of their choice; (see sample letter Figure 9-1).

NOTE: U.S. identification numbers do not exceed 5 symbols in addition to the prefix "N". These symbols may be all numbers (N55555), one to four numbers and a suffix letter (N5555A or N5A), or one to three numbers and two suffix letters (N555AB).

How to Obtain A Special Number

If you prefer a number of your own choosing or a smaller number, you may be able to obtain the exact number you want by asking the FAA Registry to assign to you a specific number of your own choice.

It would be best, however, that you list at least five numbers in case your first choice is not available. A special number of your own choosing will cost $10.00 more and you should enclose that fee with your letter. (See Sample Letter Figure 9-2).

9-2 AFFIDAVIT OF OWNERSHIP

Enclose the following additional information with your letter to the FAA Registry:

a. Class of aircraft: (airplane, glider, etc.)
b. Serial number: (Make one up if one didn't come with the plans).
c. Number of seats.
d. Type engine installed: (reciprocating, turbojet, etc.)
e. Number of engines installed: Engine make, Engine Model, Engine Serial Number.
f. State whether aircraft is built for land or sea.
g. Enclose notorized affidavit of ownership.
h. State that the aircraft was built from parts and that you are the owner. If the aircraft is built from a materials kit, a bill of sale from the kit manufacturer is also needed.
i. Type aircraft: example - Pitts, Acro Sport, etc. Not Jones- Pitts, etc.

NOTE: All of the above listed additional information required by the FAA Registry may be furnished by simply completing an Affidavit of Ownership For Amateur-Built Aircraft similar to the one illustrated in Figure 9-3. The model and serial number for your aircraft may be furnished by the designer (seller) of your plans, or it can be one made up by you.

The Affidavit must be notorized as it establishes your ownership to the airplane even though you know you did build it. It will be used by the FAA to create a file on your aircraft and will serve as a legal document and a substitute for the Bill of Sale (AC Form 8050-2) that a buyer gets when he buys any existing airplane.

Your letter requesting an identification number and the Affidavit should be both mailed to:

FAA Aircraft Registry
Department of Transportation
P.O. Box 25504
Oklahoma City, Oklahoma 73125.

In the event you would like to hand carry the information personally, their street address is 6400 S. McArthur Blvd., in Oklahoma City.

9-3 STEP 2 (REGISTERING YOUR AIRCRAFT)

After you have written the FAA Aircraft Registry, you will receive a form letter giving your Number assignment, a blank Aircraft Registration Application (AC Form 8050-1) and other registration information, see examples illustrated in Figures 9-4 & 9-5.

This letter (Figure 9-4) will have the appropriate boxes checked to indicate to you the items applicable in your case. Complete the enclosed Application for Aircraft Registration, (Figure 9-5) and return it to the FAA Aircraft Registry making sure that you have completed all items applicable. Read all the instructions carefully and comply with them otherwise your application may be returned.

Retain the **Pink** copy of the Registration Application and mail both the **White** original and the **Green** copy. Your pink copy is your authority to operate the aircraft, **when carried in the aircraft with an appropriate and current airworthiness certificate.**

9-4 WHEN TO OBTAIN YOUR NUMBER

There is no need to obtain an Identification number too soon after starting a project. This is especially true if you intend to obtain a special number of your choice. A special number costs $10.00 and will be reserved for no more than one year. You can reserve it each year thereafter for an additional $10.00 per year, however you need to contact them on this as they will **not** notify you. They automatically take that number and put it back into the pot and will give it out to anybody if you don't send them the yearly fee. If your number has not been affixed to the fuselage within this time and the registration completed, it will be necessary to pay an additional $10.00 to reserve that number for another year.

9-5 DURATION OF REGISTRATION CERTIFICATE (Part 47.41)

A certificate of Aircraft Registration as issued by the FAA is effective, unless suspended or revoked until:

a. The registration is cancelled at the written request of the holder of the Certificate.

b. The Aircraft is totally destroyed or scrapped.

c. The ownership of the aircraft is transferred.

d. The holder of the Certificate loses his U.S. citizenship.

e. Thirty days have lapsed since the death of the holder of the certificate.

f. The aircraft is registered under the laws of a foreign country (subject to international agreement.)

9-6 CHANGE OF ADDRESS

Within thirty days after any change in your mailing address, you as the holder of a Certificate of Aircraft Registration for an aircraft shall notify the FAA Aircraft Registry of your new address. A revised Certificate of Aircraft Registration is then issued, without charge.

FAA Aircraft Registry
Department of Transportation
P.O. Box 25504
Oklahoma City, Oklahoma 73125

Gentlemen:

I request that a U.S. identification number of your own choice be assigned to my Amateur-built aircraft.

An affidavit of ownership and description of the aircraft is enclosed. This aircaft has not previously been registered anywhere.

Sincerely,
R. Q. Amateur

Figure 9-1
SAMPLE LETTER:
Requesting an Identification Number

CAUTION: Registration alone does not authorize you to fly your aircraft. The aircraft must, after it has been properly registered, also obtain an Airworthiness Inspection by an inspector of the FAA, at which time the necessary Airworthiness Certificate may be issued. Then, and only then, is your aircraft ready for flight. See Figure 9-6.

These are some of the FAA forms you may need:

1. Bill of Sale AC8050-2 (if you buy an airplane).
2. Aircraft Registration Application 8050-1.
3. Application for Airworthiness certificate 8130-6.
4. Eligibility Statement 8130-12.
5. Affadavit of Ownership 8050-88.
6. Repairman Application 8610-2.
7. Bill of Sale from the kit company (if you have a kit-built aircraft this must be sent in with the registration).

When you register the aircraft:

1. Registration Application
2. Affadavit of Ownership
3. Request for Special N number (if any).
4. Bill of Sale from the kit company.

When the inspector comes:

1. Application for airworthiness
2. Eligibility statement
3. Repairman application

The above are available from the FAA office.

The a.c. registration requires a $5.00 fee.

FAA Aircraft Registry
Department of Transportation
P.O. Box 25504
Oklahoma City, Oklahoma 73125

Gentlemen:

I request that a Special identification number be assigned to my amateur-built aircraft. This aircraft has not previously been registered anywhere.

Any one of the following numbers would be acceptable:

1. N. . . . .
2. N. . . . .
3. N. . . . .
4. N. . . . .
5. N. . . . .

Enclosed is my check for the $10.00 and an affidavit of ownership which describes the aircraft.

Sincerely,
R. Q. Amateur

Figure 9-2
SAMPLE LETTER:
Requesting A Special Identification Number

AFFIDAVIT OF OWNERSHIP FOR AMATEUR-BUILT AIRCRAFT

U.S. Identification Number ______________________________

Builder's Name ______________________________

Model ______________________________

Serial Number (required) ______________________________

Class (airplane, rotorcraft, glider, etc.) ______________________________

Type of Engine Installed (reciprocating, turbopropeller, etc.)

____

Number of Engines Installed ______________________________

Manufacturer, Model, and Serial Number of each Engine Installed __________

____

____

Built for Land or Water Operation ______________________________

Number of Seats ______________________________

The above described aircraft was built from parts and I am the owner.

Address ______________________________

City ____________________ State ______________ Zip Code __________

Telephone: Home (   )_______________ Work (   )_______________

____
(Signature of Owner)

State of ____________________

County of ____________________

Subscribed and sworn to before me this _____ day of _____________, 19____.

My commission expires ______________________.

____
(Signature of Notary Public)

AC Form 8050-88 (6-87)(0052-00-559-0003) Supersedes previous edition

Figure 9-3

US Department
of Transportation
**Federal Aviation
Administration**

Mike Monroney
Aeronautical Center

P.O. Box 25504
Oklahoma City, Oklahoma 73125

Identification Number Assignment and Registration of

Amateur-Built Aircraft:______________________________

U.S. identification number N______________ has been ☐ ASSIGNED ☐ RESERVED.

A U.S. identification number of our choice may be assigned, with a $5 fee, to your amateur-built aircraft when you submit a complete description of the aircraft. The form on the back of this letter may be used, as it meets our requirements for both description and registration purposes. Authority to use a number assigned free of charge expires 90 days after the date it is issued unless the aircraft is registered within that period.

A U.S. identification number of your choice may be reserved, if available, for one year by sending a written request and a $10 fee for each number to be reserved. Please list 5 numbers, in order of preference, in case your first choice is not available. If the number is not assigned to an aircraft prior to the end of the year, the reservation will expire, but may be renewed from year to year upon request and payment of a $10 renewal fee.

NOTICE: This number may not be assigned or painted on an aircraft until approval is received from this office.

Your written request to assign the reserved number to a particular aircraft must include a complete description of the aircraft. The form on the back of this letter may be used.

The items checked below are required to complete registration of your amateur-built aircraft:

☐ Completed and signed Aircraft Registration Application; form is enclosed.

☐ Registration fee of $5.

☐ Affidavit of ownership, signed before a notary public, and showing a description of the aircraft. The form on the back of this letter meets our requirements and may be used if you wish.

Records/Conveyances Examiner

AC Form 8050-88 (6-87)(0052-00-559-0003) Supersedes previous edition

Figure 9-4

FORM APPROVED
OMB NO. 2120-0029
EXP. DATE 10/31/84

UNITED STATES OF AMERICA DEPARTMENT OF TRANSPORTATION
FEDERAL AVIATION ADMINISTRATION-MIKE MONRONEY AERONAUTICAL CENTER
AIRCRAFT REGISTRATION APPLICATION

CERT. ISSUE DATE

UNITED STATES REGISTRATION NUMBER **N**

AIRCRAFT MANUFACTURER & MODEL

AIRCRAFT SERIAL No.

FOR FAA USE ONLY

TYPE OF REGISTRATION (Check one box)

☐ 1. Individual ☐ 2. Partnership ☐ 3. Corporation ☐ 4. Co-owner ☐ 5. Gov't. ☐ 8. Foreign-owned Corporation

NAME OF APPLICANT (Person(s) shown on evidence of ownership. If individual, give last name, first name, and middle initial.)

TELEPHONE NUMBER: ( ) –

ADDRESS (Permanent mailing address for first applicant listed.)

Number and street: ______

Rural Route: P.O. Box:

CITY	STATE	ZIP CODE

☐ **CHECK HERE IF YOU ARE ONLY REPORTING A CHANGE OF ADDRESS**

ATTENTION! Read the following statement before signing this application.

A false or dishonest answer to any question in this application may be grounds for punishment by fine and / or imprisonment (U.S. Code, Title 18, Sec. 1001).

CERTIFICATION

I/WE CERTIFY:

(1) That the above aircraft is owned by the undersigned applicant, who is a citizen (including corporations) of the United States.

(For voting trust, give name of trustee: ______), or:

CHECK ONE AS APPROPRIATE:

a. ☐ A resident alien, with alien registration (Form 1-151 or Form 1-551) No. ______

b. ☐ A foreign-owned corporation organized and doing business under the laws of (state or possession) ______, and said aircraft is based and primarily used in the United States. Records of flight hours are available for inspection at ______

(2) That the aircraft is not registered under the laws of any foreign country; and
(3) That legal evidence of ownership is attached or has been filed with the Federal Aviation Administration.

NOTE: If executed for co-ownership all applicants must sign. Use reverse side if necessary.

TYPE OR PRINT NAME BELOW SIGNATURE

EACH PART OF THIS APPLICATION MUST BE SIGNED IN INK.	SIGNATURE	TITLE	DATE
	SIGNATURE	TITLE	DATE
	SIGNATURE	TITLE	DATE

NOTE: Pending receipt of the Certificate of Aircraft Registration, the aircraft may be operated for a period not in excess of 90 days, during which time the PINK copy of this application must be carried in the aircraft.

AC FORM 8050-1 (1-83) (0052-00-628-9005)

Figure 9-5

If it is built by more than one person you will be better off putting on all signatures. The builder's name refers to the person who built the **major** portion of the aircraft. The main builder goes on the data plate. If the aircraft wasn't built from parts by one person, (if you bought a project ready to fly - unregistered) the builder's name should be in the signature of owner/builder and you should cross out the words "by the undersigned". It is important that it not be falsified because it opens you to a fine and imprisonment.

Form Approved
O.M.B. NO. 2120-0018

US Department of Transportation
Federal Aviation Administration

ELIGIBILITY STATEMENT
AMATEUR-BUILT AIRCRAFT

Instructions: Print or type all information except signature. Submit original to an authorized FAA representative. Applicant completes Section I thru III. Notary Public Completes Section IV.

I. REGISTERED OWNER INFORMATION

Name(s) ______________________________

Address(es) ______________________________
No. & Street — City — State — Zip

Telephone No.(s) ( ) ______________ Residence ( ) ______________ Business

II. AIRCRAFT INFORMATION

Model ______________ Engine(s) Make ______________

Assigned Serial No. ______________ Engine(s) Serial No.(s) ______________

Registration No. ______________ Prop./Rotor(s) Make ______________

Aircraft Fabricated: Plan ☐ Kit ☐ Prop./Rotor(s) Serial No.(s) ______________

III. MAJOR PORTION ELIGIBILITY STATMENT OF APPLICANT

I certify the aircraft identified in Section II above was fabricated and assembled by ______________________________
Name of Person(s) (Please Print)
for my (their) education or recreation. I (we) have records to support this statement and will make them available to the FAA upon request.

— NOTICE —

Whoever in any matter within the jurisdiction of any department or agency of the United States knowingly and willfully falsifies, conceals or covers up by any trick, scheme, or device a material fact, or who makes any false, fictitious or fraudulent statements or representations, or makes or uses any false writing or document knowing the same to contain any false, fictitious or fraudulent statement or entry, shall be fined not more than $10,000 or imprisoned not more than 5 years, or both (U.S. Code, Title 18, Sec. 1001.)

APPLICANT'S DECLARATION

I hereby certify that all statements and answers provided by me in this statement form are complete and true to the best of my knowledge, and I agree that they are to be considered part of the basis for issuance of any FAA certificate to me. I have also read and understand the Privacy Act statement that accompanies this form.

Signature of Applicant *(In Ink)*	Date

IV. NOTARIZATION STATEMENT

FAA Form 8130-12 (4-89)

Figure 9-6

SECTION 10

Affixing Nationality And Registration Marks

10-1 DISPLAY OF MARKS (Reference is FAR Part 45.23)

After you obtain the registration for your aircraft, the Registration numbers or marks must be affixed to the aircraft in some permanent fashion. The marks must be legible and have no ornamentation. They must contrast in color with the background.

The marks displayed on the aircraft shall include the letter "N" signifying U.S. Registry, followed by the registration number issued for the aircraft.

In addition, amateur-built (Experimental) aircraft must have displayed on that aircraft near each entrance to the cabin or cockpit, in letters not less than 2" nor more than 6" in height, the word, "EXPERIMENTAL".

10-2 LOCATION OF MARKS ON FIXED WING AIRCRAFT (Reference is FAR Part 45.25)

(b)(1) If displayed on the vertical tail surfaces, horizontally on both surfaces, horizontally on both surfaces of a single vertical tail or on the outer surfaces of a multivertical tail. However, an aircraft on which marks at least 3" high may be displayed and in accordance with 45.29(b)(1), the marks may be displayed vertically on the vertical tail surfaces.

(2) If displayed on the fuselage surfaces, horizontally on both sides of the fuselage between the trailing edge of the wing and the leading edge of the horizontal stabilizer. However, if engine pods or other appurtenances are located in this area and are an integral part of the fuselage side surfaces, the operator may place the marks on those pods or appurtenances.

10-3 SIZE OF MARKS

FAR 45.29(b)(1)(iii) states "Marks at least 3" high may be displayed on an aircraft for which an experimental certificate has been issued under 21.191(d) or 21.191(g) for operating as an exhibition aircraft or as an amateur-built aircraft when the maximum cruising speed of the aircraft does not exceed 180 knots Calibrated Air Speed (CAS).

And (c) characters must be two-thirds as wide as they are high except "1" which must be one-sixth as wide as it is high and the letters "M" and "W" which may be as wide as they are high.

(d) Characters must be formed by solid lines one-sixth as thick as the character is high.

(e) Spacing. The space between each character may not be less than one-fourth of the character width.

10-4 EXHIBITION, ANTIQUE AND OTHER AIRCRAFT... SPECIAL RULES APPLY

FAR 45.22 states the following: (b) A small U.S. registered aircraft built at least 30 years ago or a U.S. registered aircraft for which an experimental certificate has been issued under 21.191(d) or 21.191(g) for operating as an exhibition aircraft or as an amateur-built aircraft and which has the same external configuration as an aircraft built at least 30 years ago, may be operated without displaying marks in accordance with 45.21 and 45.23 through 45.33 if:

(1) It displays in accordance with 45.21(c) marks at least 2" high on each side of the fuselage or vertical tail surface consisting of the Roman capital letter "N" followed by (ii) a symbol appropriate to the airworthiness certificate of the aircraft, i.e. "X" experimental followed by the U.S. registration number of the aircraft.

This is the end of the comments from the FAR's. What this means is that if the prototype of the aircraft that you are building was flown at least 30 years ago, that you may use 2" numbers preceded by an "NX" or an "N". This would be appropriate for the Pitts Special and other amateur-built aircraft if the prototype was completed over 30 years ago. Otherwise, the amateur-builder must use 3" high numbers if maximum cruise is under 180 knots.

(2) 12" numbers must be displayed on the fuselage surfaces, horizontally on both sides of the fuselage between the trailing edge of the wing and the leading edge of the horizontal stabilizer. 2" and 3" numbers may be placed horizontally on the vertical tail. 2" numbers may be placed under the horizontal stabilizer.

FIGURE 10-1 METAL IDENTIFICATION PLATES
The plate is covered by FAR 45.11 and 45.13.

10-5 IDENTIFICATION PLATE (Reference is FAR Part 45.11)

In addition to affixing the aircraft's registration number to the sides of the fuselage, the builder must also identify his aircraft by attaching an identification plate to the aircraft's structure.

The identification data required to be inscribed on the plate for amateur-built aircraft shall include the following information:

a. Builder's name and address: John Amateur
123 Wright St.
Anytown, USA

b. Model designation: Pitts S1C

c. Builder's serial number: 01 (pick one you like)

d. Date of manufacture: 01/01/91

The identification plate containing these essential elements must be of fireproof material and must be secured in such a manner that it will not likely be defaced or removed during normal service, or lost or destroyed in an accident. It must be secured to the aircraft at an accessible location adjacent to and aft of the rearmost entrance door or on the fuselage near the tail surfaces. It must be legible to a person on the ground.

The identification plate information must be marked thereon by etching, stamping, engraving, or other acceptable fireproof marking.

Metal plates which comply with these requirements may be purchased from the Experimental Aircraft Association for the very nominal fee of $5.00 per each set of two as illustrated. (See Figure 10-1)

10-6 Location of Registration Marks for Rotorcraft, Airships and Balloons is specified in FAR 45.27.

Markings on various homebuilt aircraft — as can be seen, amateur builders find ways of being creative and incorporating markings in the overall paint scheme of their aircraft.

The Airworthiness Certificate . . . How To Get It

11-1 TYPE OF AIRWORTHINESS CERTIFICATE TO BE ISSUED

Any owner of an aircraft that has been properly registered may apply for the issue of an airworthiness certificate. Without an airworthiness certificate an aircraft may not be flown legally.

Amateur-built aircraft are currently certificated in the special airworthiness category and are issued experimental certificates.

11-2 AMENDMENT OR MODIFICATION

No airworthiness certificate may be amended or modified without formal application to and approval from the FAA.

11-3 DURATION OF THE AIRWORTHINESS CERTIFICATE

Unless it is sooner surrendered, suspended, revoked or a shorter termination date was otherwise established by the FAA, an experimental airworthiness certificate is effective for an indefinite period. However, the FAA can reinspect an experimental aircraft at any time and amend, modify, suspend or revoke the certificate.

11-4 PROCEDURE FOR THE ISSUANCE OF THE ORIGINAL AIRWORTHINESS CERTIFICATE (The Reference is FAR 21.193)

The final step toward obtaining your airworthiness certificate is your written request to the FAA for a Pre-Certification Inspection. See Figure 11-1.

At this point there appears to be considerable difference in the formal procedures required by the different FAA offices. When all has been said and done, however, the FAA will have essentially the same information for their record. The difference being that some FAA Inspectors will help the individual builder with his paperwork at the time of inspection. This may even include the completion of an application for Airworthiness Certificate, FAA Form 8130-6, (See Figure 11-2), the test area request, and the other information required by FAR 21.193. Regardless, by this time you should have learned what is expected of you by the FAA Inspector who has been checking your project and you should be guided accordingly.

If for some reason you did not ask your Inspector what would be expected of you, or if you are newly moved into the area, it would be best to submit with your letter the following information as prescribed by FAR 21.193.

Eligibility Statement - Amateur-Built Aircraft - Privacy Act, FAA Form 8130.

1. A statement giving the purpose for which the aircraft is to be used. (Recreation, education, testing and experimentation).

2. The estimated time or number of flights required for the experiment. (Usually 25 hours for aircraft equipped with certified aircraft engine and propeller combinations and 40 hours for those with non-aircraft type engines propeller combinations installed.)

3. The areas over which the experiment will be conducted. (Request an area encompassing a 25 mile radius for day VFR operations. Exclude congested areas and airways but try to include nearby airports even if a few miles beyond the 25 mile radius).

4. Include a three-view drawing of the aircraft, or photos — top-side — front view. The drawing should be neat and in proportion but need not be of a professional nature. The following data should also be included or shown on the three-view drawing: (See Figure 11-3).

5. Horsepower rating of the engine and type of propeller.

6. Empty weight and maximum weight at which aircraft will be operated.

7. Number of seats installed and their arrangement with respect to each other.

8. Whether single or dual controlled.

9. Fuel and oil capacities.

10. Maximum speed at which the applicant expects to operate the aircraft.

After all this information has been sent to your nearest District Office you will receive instructions regarding the time and date for the inspection.

> 718 Homebuilder St.
> Hometown, WI
> (Date)
>
> Department of Transportation
> Federal Aviation Administration
> (Address)
> (City & State)
>
> Gentlemen:
>
> I submit this application for N ---, Acro Sport aircraft, for an airworthiness certificate in the experimental category, amateur built aircraft, for the purpose of recreation, education, testing and other experimentation flying in accordance with FAR 21.193. Forms 8130-6 and 8130-12 enclosed.
>
> Flights are to be conducted in the assigned flight test area.
>
> Sincerely,
> Roland Q. Amateur

Figure 11-1

SAMPLE LETTER: Application for Origional Airworthiness Certificate

NOTE: Enclose FAA Form 8130-6

Form Approved
O.M.B. No. 2120-0018

US Department of Transportation
Federal Aviation Administration

APPLICATION FOR AIRWORTHINESS CERTIFICATE

INSTRUCTIONS — Print or type. Do not write in shaded areas; these are for FAA use only. Submit original only to an authorized FAA Representative. If additional space is required, use an attachment. For special flight permits complete Sections II and VI or VII as applicable.

I. AIRCRAFT DESCRIPTION

1. REGISTRATION MARK	2. AIRCRAFT BUILDER'S NAME *(Make)*	3. AIRCRAFT MODEL DESIGNATION	4. YR. MFR.	FAA CODING
N123	I. M. Builder	Lark	1991	
5. AIRCRAFT SERIAL NO.	6. ENGINE BUILDER'S NAME *(Make)*	7. ENGINE MODEL DESIGNATION		
L-1	Lycoming	0-320		
8. NUMBER OF ENGINES	9. PROPELLER BUILDER'S NAME *(Make)*	10. PROPELLER MODEL DESIGNATION	11. AIRCRAFT IS *(Check if applicable)*	
1	Warnke	Model 1	IMPORT	

II. CERTIFICATION REQUESTED

APPLICATION IS HEREBY MADE FOR: *(Check applicable items)*

A 1 STANDARD AIRWORTHINESS CERTIFICATE *(Indicate category)* — NORMAL / UTILITY / ACROBATIC / TRANSPORT / GLIDER / BALLOON

B SPECIAL AIRWORTHINESS CERTIFICATE *(Check appropriate items)*

- 2 LIMITED
- 5 PROVISIONAL *(Indicate class)*: 1 CLASS I; 2 CLASS II
- 3 RESTRICTED *(Indicate operation(s) to be conducted)*: 1 AGRICULTURE AND PEST CONTROL; 2 AERIAL SURVEYING; 3 AERIAL ADVERTISING; 4 FOREST *(Wildlife conservation)*; 5 PATROLLING; 6 WEATHER CONTROL; 7 CARRIAGE OF CARGO; 0 OTHER *(Specify)*
- 4 [X] EXPERIMENTAL *(Indicate operation(s) to be conducted)*: 1 RESEARCH AND DEVELOPMENT; 2 AMATEUR BUILT; 3 EXHIBITION; 4 RACING; 5 CREW TRAINING; MKT. SURVEY; 0 TO SHOW COMPLIANCE WITH FAR
- 8 SPECIAL FLIGHT PERMIT *(Indicate operation to be conducted, then complete Section VI or VII as applicable on reverse side)*: 1 FERRY FLIGHT FOR REPAIRS, ALTERATIONS, MAINTENANCE OR STORAGE; 2 EVACUATE FROM AREA OF IMPENDING DANGER; 3 OPERATION IN EXCESS OF MAXIMUM CERTIFICATED TAKE-OFF WEIGHT; 4 DELIVERING OR EXPORT; 5 PRODUCTION FLIGHT TESTING; 6 CUSTOMER DEMONSTRATION FLIGHTS

C 6 MULTIPLE AIRWORTHINESS CERTIFICATE *(Check ABOVE "Restricted Operation" and "Standard" or "Limited," as applicable.)*

III. OWNER'S CERTIFICATION

A. REGISTERED OWNER *(As shown on certificate of aircraft registration)* — IF DEALER, CHECK HERE ⟶

NAME	ADDRESS
I. M. Builder	2 Main Street Anytown, U S A

B. AIRCRAFT CERTIFICATION BASIS *(Check applicable blocks and complete items as indicated)*

- AIRCRAFT SPECIFICATION OR TYPE CERTIFICATE DATA SHEET *(Give No. and Revision No.)*
- AIRWORTHINESS DIRECTIVES *(Check if all applicable AD's complied with and give latest AD No.)*
- AIRCRAFT LISTING *(Give page number(s))*
- SUPPLEMENTAL TYPE CERTIFICATE *(List number of each STC incorporated)*

C. AIRCRAFT OPERATION AND MAINTENANCE RECORDS

CHECK IF RECORDS IN COMPLIANCE WITH FAR 91.173	TOTAL AIRFRAME HOURS	3 EXPERIMENTAL ONLY *(Enter hours flown since last certificate issued or renewed)*

D. CERTIFICATION — I hereby certify that I am the registered owner (or his agent) of the aircraft described above, that the aircraft is registered with the Federal Aviation Administration in accordance with Section 501 of the Federal Aviation Act of 1958, and applicable Federal Aviation Regulations, and that the aircraft has been inspected and is airworthy and eligible for the airworthiness certificate requested.

DATE OF APPLICATION	NAME AND TITLE *(Print or type)*	SIGNATURE
12-9-91	I. M. Builder	

IV. INSPECTION AGENCY VERIFICATION

A. THE AIRCRAFT DESCRIBED ABOVE HAS BEEN INSPECTED AND FOUND AIRWORTHY BY: *(Complete this section only if FAR 21.183(d) applies)*

- 2 FAR PART 121 OR 127 CERTIFICATE HOLDER *(Give Certificate No.)*
- 3 CERTIFICATED MECHANIC *(Give Certificate No.)*
- 6 CERTIFICATED REPAIR STATION *(Give Certificate No.)*
- 5 AIRCRAFT MANUFACTURER *(Give name of firm)*

DATE	TITLE	SIGNATURE

V. FAA REPRESENTATIVE CERTIFICATION

(Check ALL applicable blocks in items A and B)

A. I find that the aircraft described in Section I or VII meets requirements for: THE CERTIFICATE REQUESTED; 4 AMENDMENT OR MODIFICATION OF CURRENT AIRWORTHINESS CERTIFICATE

B. Inspection for a special flight permit under Section VII was conducted by: FAA INSPECTOR; FAA DESIGNEE; CERTIFICATE HOLDER UNDER: FAR 65; FAR 121, 127 or 135; FAR 145

DATE	DISTRICT OFFICE	4 DESIGNEE'S SIGNATURE AND NO.	1 FAA INSPECTOR'S SIGNATURE

FAA Form 8130-6 (11-88) SUPERSEDES PREVIOUS EDITION

Figure 11-2 (see reverse side)

*U.S.GPO:1989-241-672/90913

VI. PRODUCTION FLIGHT TESTING

A. MANUFACTURER

NAME	ADDRESS

B. PRODUCTION BASIS *(Check applicable item)*

- [ ] PRODUCTION CERTIFICATE *(Give production certificate number)*
- [ ] TYPE CERTIFICATE ONLY
- [ ] APPROVED PRODUCTION INSPECTION SYSTEM

C. GIVE QUANTITY OF CERTIFICATES REQUIRED FOR OPERATING NEEDS →

DATE OF APPLICATION	NAME AND TITLE *(Print or type)*	SIGNATURE

VII. SPECIAL FLIGHT PERMIT PURPOSES OTHER THAN PRODUCTION FLIGHT TEST

A. DESCRIPTION OF AIRCRAFT

REGISTERED OWNER	ADDRESS
BUILDER *(Make)*	MODEL
SERIAL NUMBER	REGISTRATION MARK

B. DESCRIPTION OF FLIGHT CUSTOMER DEMONSTRATION FLIGHTS ☐ *(Check if applicable)*

FROM	TO	
VIA	DEPARTURE DATE	DURATION

C. CREW REQUIRED TO OPERATE THE AIRCRAFT AND ITS EQUIPMENT

☐ PILOT ☐ CO-PILOT ☐ NAVIGATOR ☐ OTHER *(Specify)*

D. THE AIRCRAFT DOES NOT MEET THE APPLICABLE AIRWORTHINESS REQUIREMENTS AS FOLLOWS:

NOT APPLICABLE

E. THE FOLLOWING RESTRICTIONS ARE CONSIDERED NECESSARY FOR SAFE OPERATION *(Use attachment if necessary)*

F. CERTIFICATION — I hereby certify that I am the registered owner (or his agent) of the aircraft described above; that the aircraft is registered with the Federal Aviation Administration in accordance with Section 501 of the Federal Aviation Act of 1958, and applicable Federal Aviation Regulations; and that the aircraft has been inspected and is airworthy for the flight described.

DATE	NAME AND TITLE *(Print or type)*	SIGNATURE
12-9-91	I. M. Builder	

VIII. AIRWORTHINESS DOCUMENTATION (FAA use only)

☐	**A.** Operating Limitations and Markings in Compliance with FAR 91.31 as Applicable	☐	**G.** Statement of Conformity, FAA Form 8130-9 *(Attach when required)*
☐	**B.** Current Operating Limitations Attached	☐	**H.** Foreign Airworthiness Certification for Import Aircraft *(Attach when required)*
☐	**C.** Data, Drawings, Photographs, etc. *(Attach when required)*	☐	**I.** Previous Airworthiness Certificate Issued in Accordance with
☐	**D.** Current Weight and Balance Information Available in Aircraft		**FAR** ______ **CAR** ______ *(Original Attached)*
☐	**E.** Major Repair and Alteration, FAA Form 337 *(Attach when required)*	☐	**J.** Current Airworthiness Certificate Issued in Accordance with
☐	**F.** This Inspection Recorded in Aircraft Records		**FAR** ______ *(Copy attached)*

Figure 11-2

ACRO II SPECIFICATIONS

Upper Wing Span	21'8"
Lower Wing Span	20'9"
Wing Area	152 sq. ft.
Length	18'10¼"
Height	79¾"
Powerplant	Lycoming 180 hp
Can by powered with engines of 115 hp	
Landing Gear Tread	6'¾"
Empty Weight	875 lbs.
Gross Weight	1520 lbs.
Fuel Capacity	26 gals.
Baggage	30 lbs.
Chord	43"
Wing Airfoil	M-6
Cruise . . (180 hp Lyc.)	123 mph
Stall Speed	53 mph

Rigging	
Incidence	
Upper Wing	+1½°
Lower Wing	+1½°
Dihedral	
Upper Wing	0°
Lower Wing	+2°
Aileron, Up and Down Travel	25°
Rudder, Left and Right Travel	32°
Elevator, Up and Down Travel	30°
Tire Pressure	27-28 lbs.

EMPTY WEIGHT C.G. ON THE PROTOTYPE AIRCRAFT WAS 8" FORWARD OF THE LE OF THE LOWER WING.

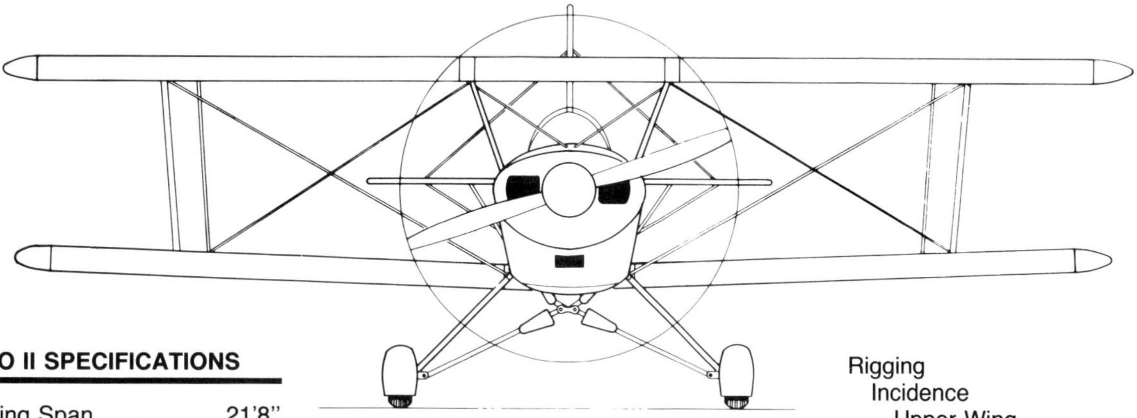

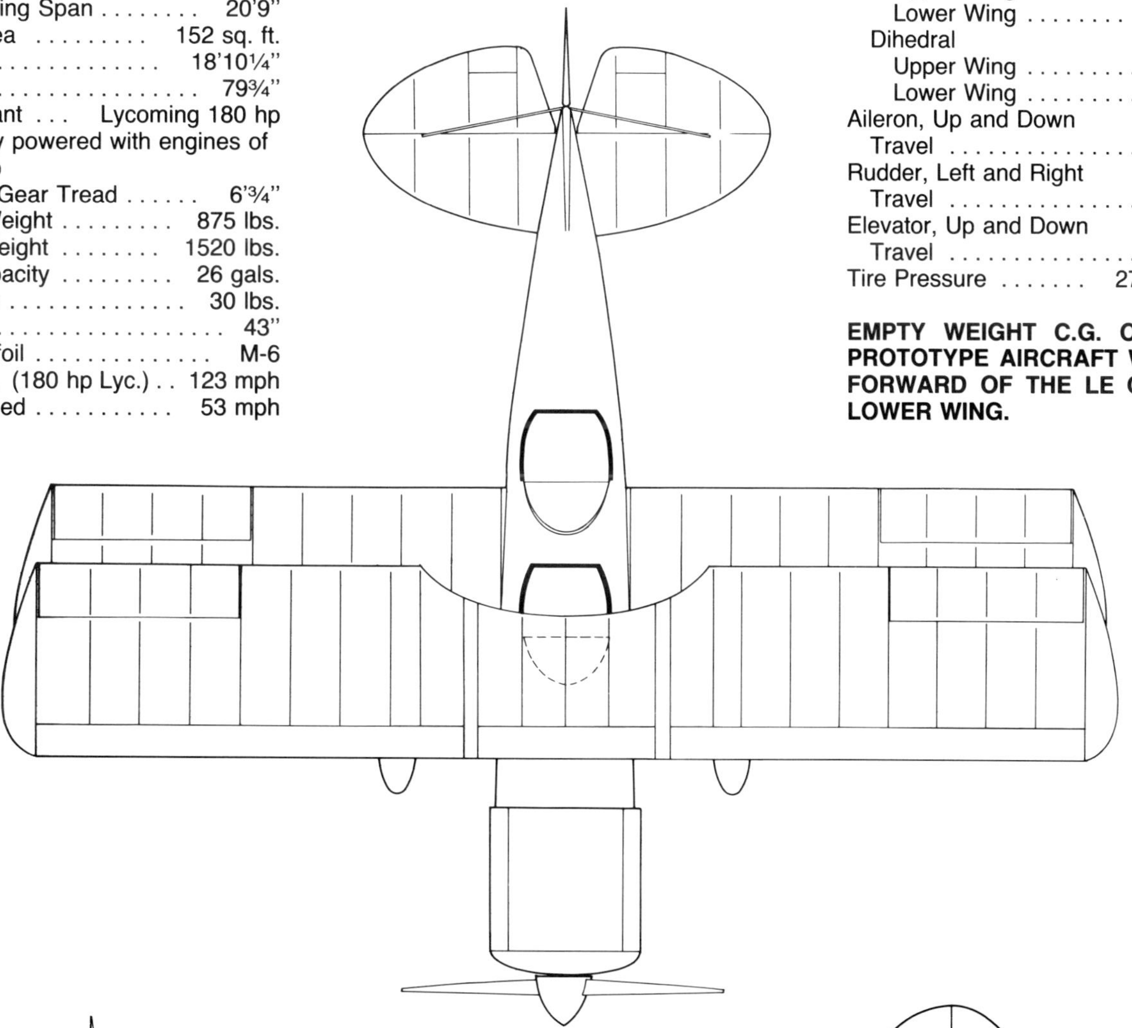

Figure 11-3

FORM APPROVED
OMB NO. 2120-0042

UNITED STATES OF AMERICA
DEPARTMENT OF TRANSPORTATION FEDERAL AVIATION ADMINISTRATION

AIRCRAFT BILL OF SALE

FOR AND IN CONSIDERATION OF $ THE UNDERSIGNED OWNER(S) OF THE FULL LEGAL AND BENEFICIAL TITLE OF THE AIRCRAFT DESCRIBED AS FOLLOWS:

UNITED STATES REGISTRATION NUMBER **N**

AIRCRAFT MANUFACTURER & MODEL

AIRCRAFT SERIAL No.

DOES THIS DAY OF 19
HEREBY SELL, GRANT, TRANSFER AND DELIVER ALL RIGHTS, TITLE, AND INTERESTS IN AND TO SUCH AIRCRAFT UNTO:

Do Not Write In This Block
FOR FAA USE ONLY

PURCHASER

NAME AND ADDRESS
(IF INDIVIDUAL(S), GIVE LAST NAME, FIRST NAME, AND MIDDLE INITIAL.)

DEALER CERTIFICATE NUMBER

AND TO EXECUTORS, ADMINISTRATORS, AND ASSIGNS TO HAVE AND TO HOLD SINGULARLY THE SAID AIRCRAFT FOREVER, AND WARRANTS THE TITLE THEREOF.

IN TESTIMONY WHEREOF HAVE SET HAND AND SEAL THIS DAY OF 19

SELLER

NAME (S) OF SELLER (TYPED OR PRINTED)	SIGNATURE (S) (IN INK) (IF EXECUTED FOR CO-OWNERSHIP, ALL MUST SIGN.)	TITLE (TYPED OR PRINTED)

ACKNOWLEDGMENT (NOT REQUIRED FOR PURPOSES OF FAA RECORDING: HOWEVER, MAY BE REQUIRED BY LOCAL LAW FOR VALIDITY OF THE INSTRUMENT.)

ORIGINAL: TO FAA

AC FORM 8050-2 (8-85) (0052-00-629-0002)

Figure 11-4

11-5 SELLING YOUR AIRCRAFT

You must make an Aircraft Bill of Sale, (See Figure 11-4), for the purchaser of your aircraft. Unless entries or notations on your certificate state to the contrary, no new certificate must be obtained prior to the normal expiration date on the Airworthiness Certificate.

11-6 MAJOR AND MINOR CHANGES

Changes or modifications to an aircraft are classified as minor or major. A change is minor if it has no appreciable effect on the weight, balance, structural integrity, operational characteristics, or other characteristics affecting the airworthiness of the aircraft. All other changes are classified as major.

The FAA inspector will usually caution the builder on handing him his new Airworthiness Certificate that any major modification or changes to the aircraft by the builder would automatically void the certificate and that the FAA would have to reinspect the aircraft. He will probably advise the builder to contact the GADO in order to discuss any proposed modification to ascertain whether or not it would affect the current Airworthiness Certificate.

See FAR Part 21.93 for definitions of major changes and alterations.

Left — Fuselage of all wood amphibian prior to skinning. Shape of bottom of hull is important as it assists aircraft in breaking water suction on takeoff and reduces takeoff distance. It also helps reduce splashing. Below — Nose rib detail showing drag and antidrag rod installation. Wing will be varnished prior to applying aluminum leading edge.

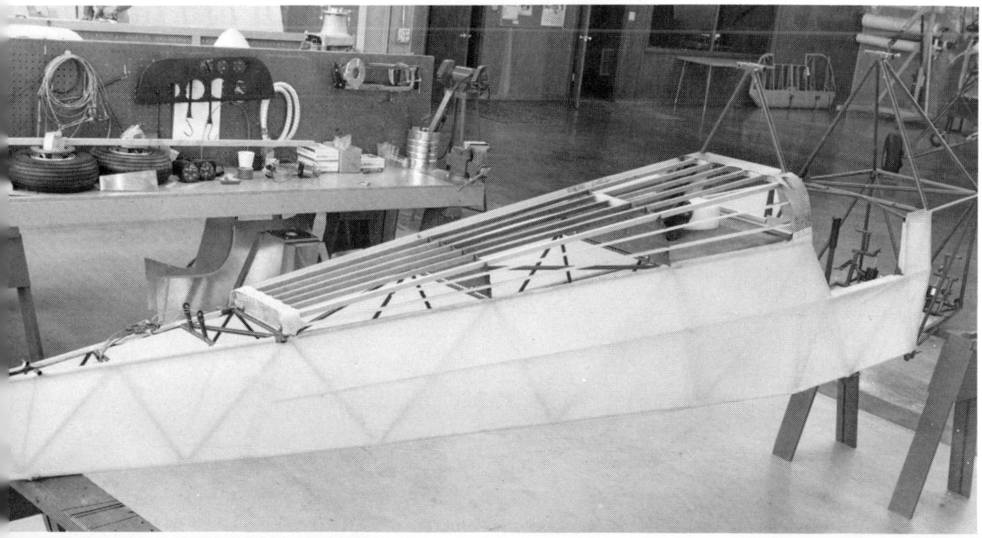

Partially covered fuselage of Pober Pixie. Most modern covering materials use heat to tighten and shrink fabric. Temperatures are no longer critical to applying dope — simplifying covering tremendously.

Rudder and hydraulic toe brake installation on the Pixie. As is evident, this was taken during the early stages of construction.

Operation Of Amateur-Built Aircraft

12-1 GENERAL

Amateur-built aircraft are governed by the same operating rules as any other aircraft. These rules are contained in FAR Part 91, "General Operating and Flight Rules." One exception is a Gyroglider which is flown while attached to a ground or water towing vehicle. It is considered a kite as long as it is under 150 feet AGL and on tow from the ground. As such, it is subject to FAR Part 101.11, 101.13, 101.15, "Moored Balloons, Kites, Unmanned Rockets, and Unmanned Free Balloons."

12-2 WHO CAN FLY A HOMEBUILT?

Pilot certificate requirements for amateur-built aircraft are fundamentally the same as certificate requirements for any other type of aircraft and are described in FAR Part 61.

12-3 WHO CAN FLY THE FIRST FLIGHT TEST?

A minimum of a private pilots license is required. Either a student pilot's license with an endorement, "safe to solo", in that homebuilt aircraft or a recreational or higher grade pilot certificate is necessary in order to fly a homebuilt solo. After the initial passenger restrictions are removed, you can carry passengers in your homebuilt if you have a recreational pilot license or above. In addition, an aircraft with an experimental certificate amateur-built may be used for fight instruction where no charges or renumeration for the use of the aircraft are involved. A pilot or owner may avail themselves to the service of an instructor to take dual instruction in an experimental aircraft. However, a commercial operator may not provide such an aircraft for the purpose of giving flight instruction for hire. In this instance, the student pilot can compensate the CFI but cannot use his aircraft for hire purposes. In addition, a student pilot who has taken flight instruction in his amateur-built aircraft cannot use this aircraft for the final test for a rating. You must present a type certificated aircraft for issuance of appropriate category and class rating in accordance with FAR 61.45, (it is up to the inspector). Occasionally you'll find that an inspector will fly in an amateur built aircraft 2-place if he likes it. A student pilot can be signed off safe to solo an amateur-built aircraft by a CFI, however not many CFIs will do this for the test period. A person who is only a student pilot can't fly the test period! There is no way that a CFI can be considered essential crew even if the owner is a student pilot and needs instruction because if there is a problem it will put two people in jeopardy instead of just one.

A holder of an airplane pilot certificate can fly a rotorcraft in solo flight, (unless the operating limitations say otherwise), without having a rotorcraft rating, although it is not recommended. He must have the rotorcraft rating in order to carry passengers, however . . . For recreational license endorsement for rotorcraft the pilot must have a total of 30 hours of flight instruction and solo flight time in aircraft. For a helicopter rating including at least 15 hours of helicopter solo flight time plus 4 hours of dual. For a gyroplane rating, including at least 10 hours of gyrocraft or solo flight time in gyroplanes plus 4 hour dual (covered under FAR-61.100).

Built from a kit this RV-6, 2-place, side by side, all-metal aircraft was built by Alan Tolle in just six months. Engine is a Lycoming 0-360 A1A taken from a 180 Comanche. Tolle had previously built a tandem RV-4.

Replicas of famous aircraft — be they glamorous racers from the golden age of air racing — World War I fighters, or distinctive vintage aircraft — are emerging from builders' shops in greater numbers each year. This Wedell-Williams Model 44 racer appeared at the EAA convention in 1990. It was built by Jim and Liz Clevenger, Marion, NC. A 450 hp P&W engine powers the red and black beauty. The original racer won the Thompson race in 1933 and also set a world speed record that year at 305.53 mph. In 1934 the aircraft also won the transcontinental Bendix race.

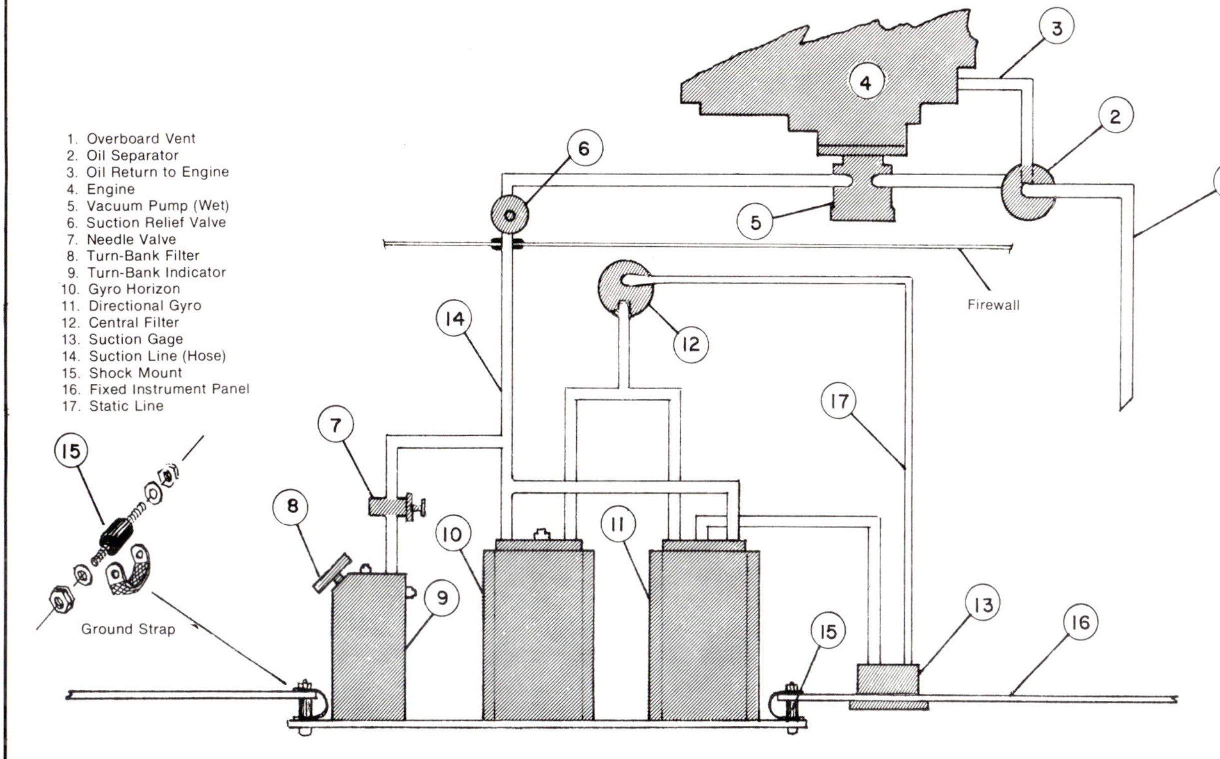

GYRO INSTRUMENT SYSTEM (SCHEMATIC)

Operating Limitations

13-1 MANDATORY TEST FLIGHT PROVING PHASE

All amateur-built sport aircraft as well as standard aircraft have federally imposed operating limitations.

Upon satisfactory completion of the necessary final FAA Inspection of the aircraft and whatever ground tests may be required, the FAA Inspector will issue your amateur-built "Experimental" Airworthiness Certificate.

He will also issue a form letter establishing the operating limitations applicable to your aircraft during its mandatory flight proving period. These Special Airworthiness Experimental Operating Limitations must be displayed in the aircraft at all times. (See sample Operating Limitations, Figure 13-1).

The operating limitations imposed on the aircraft during its flight proving period will be more stringent than those issued later after the mandatory flight testing phase has been completed.

This phase may begin with the issuance of the aircraft's initial airworthiness certificate and the original operating limitations. At this time the FAA Inspector will acquaint you with the requirement for a mandatory flight test and proving period. This flying will be confined to an assigned flight area approved by the FAA Inspector.

The presence of the FAA Inspector is not required, by regulation, at the initial flight of the experimental amateur-built aircraft. If time permits, however, it is not unusual for him to attend.

If he deems necessary, the inspector could issue a permit for a single flight within the boundaries of the airport and, upon witnessing the safe completion of the test, issue a further permit for more extended flights within the permissible area.

A tremendous responsibility for the safe operation of the experimental aircraft rests on the FAA Inspector. If the plane has any new and unusual features he will naturally tend to treat its first flights with care. Also pilot qualification and skill is a consideration.

13-2 PURPOSE OF THE FLIGHT TEST PERIOD

A flight test period is necessary to show to the FAA that the aircraft is controllable throughout its normal range of speeds and throughout all the maneuvers to be executed. It will also serve to prove that the aircraft has no hazardous operating characteristics or design features.

13-3 DURATION OF MANDATORY FLIGHT TEST PERIOD

For standard aircraft type engines: When an FAA approved aircraft engine/propeller combination is installed the flight test period is usually limited to 25 hours of flight time.

For non-aircraft type or automotive engines: An aircraft equipped with such an engine is required to be flown for a longer test period, usually at least 40 hours, to prove its reliability.

NOTE: It should be understood that the local FAA Inspector has the prime responsibility in determining the extent of the flight test period to be required for your aircraft. He is permitted to exercise considerable discretion in extending or in reducing the number of hours required to be flown during this period.

13-4 FLIGHT TEST AREA

The FAA Inspector will authorize the flight tests to be carried out in a designated and limited test area, usually, within a 25 mile radius of the aircraft's base of operations.

He will insure that the area selected is not over densely populated areas or in a congested airway.

In assigning the flight test area the FAA Inspector may modify the size and shape of the area to suit the best purposes of the flight test program. In some locations, particularly around bigger cities where air traffic is heavy, a flight test area may not be practical. The builder must be prepared to expect that an approved flight test area may not be the one chosen by him as the most convenient.

13-5 OTHER LIMITATIONS DURING THE FLIGHT TEST PERIOD

As a rule, the carrying of passengers or other crew members will not be permitted unless necessary to the safe operation of that aircraft. Flying as a student would require being signed off by a CFI. This is not a training period. The student pilot will have to hire or obtain a private pilot or a recreational pilot or above for the test period.

13-6 AIRCRAFT FLIGHT LOG

During the flight test period, the pilot should record the aircraft flight history in an appropriate log book. This should be in addition to any engine tachourmeter or engine hourmeter that may be installed in the aircraft.

Specifically, the duration of each individual flight should be recorded including the number of landings made.

A full description of any mishaps, however minor, or any experiences not entirely normal that occur during the flight experience period should also be duly recorded.

Although not required, it is strongly recommended that all operating data be recorded flight by flight. Such information as airspeeds, cylinder head temperatures, oil temperatures and pressures, altitudes and free air temperatures, etc., will be very valuable and may be used to determine or establish the various performance figures and operating characteristics of the aircraft.

Although the FAA Inspector is required by law to apply certain basic restrictions permanently to the amateur-built aircraft he is certificating, he can apply whatever other limitations he deems necessary at his own discretion. Unfortunately, nothing in the regulations states that the initial restrictions are required to be removed after successful completion of the test period . . . they only may be modified.

After the mandatory flight test period . . . then what?

APPENDIX 9. SAMPLE LIST OF OPERATING LIMITATIONS

THESE OPERATING LIMITATIONS SHALL BE ACCESSIBLE TO THE PILOT

EXPERIMENTAL OPERATING LIMITATIONS OPERATING AMATEUR-BUILT AIRCRAFT

REG. NO. ____________________ SERIAL NO. ________
MAKE ____________________ MODEL ________

Phase I, Initial Flight Test in Restricted Area:

1. No person may operate this aircraft for other than the purpose of operating amateur-built aircraft to accomplish the operation and flight test outline in the applicant's letter dated ____________ in accordance with FAR section 21.193. Phase I and II amateur-built operations shall be conducted in accordance with applicable air traffic and general operating rules of FAR Part 91 and the additional limitations herein prescribed under the provisions of FAR section 91.42 (new FAR section 91.319).

2. The initial _______ hours of flight shall be conducted within the geographical area described as follows:

____

____

3. Except for takeoffs and landings, no person may operate this aircraft over densely populated areas or in congested airways.

4. This aircraft is approved for day VFR operation only.

5. Unless prohibited by design, acrobatics are permitted in the assigned flight test area. All acrobatics are to be conducted under the provisions of FAR section 91.71 (new FAR section 91.303).

6. No person may be carried in this aircraft during flight unless that person is required for the purpose of the flight.

7. The cognizant FAA office must be notified and their response received in writing prior to flying this aircraft after incorporating a major change as defined by FAR section 21.93.

8. The operator of this aircraft shall notify the control tower of the experimental nature of this aircraft when operating into or out of airports with operating control towers.

Figure 13-1 (see reverse side)

APPENDIX 9. SAMPLE LIST OF OPERATING LIMITATIONS (CONT'D)

9. The pilot-in-command of this aircraft must, as applicable, hold an appropriate category/class rating, have an aircraft type rating, have a flight instructor's logbook endorsement or possess a "Letter of Authorization" issued by an FAA Flight Standards Operations Inspector.

10. This aircraft does not meet the requirements of the applicable, comprehensive, and detailed airworthiness code as provided by Annex 8 to the Convention on International Civil Aviation. This aircraft may not be operated over any other country without the permission of that country.

Phase II:

Following satisfactory completion of the required number of flight hours in the flight test area, the pilot shall certify in the logbook that the aircraft has been shown to comply with FAR section 91.42(b) (new FAR section 91.319). Compliance with FAR section 91.42(b) (new FAR section 91.319) shall be recorded in the aircraft logbook with the following or similarly worded statement:

"I certify that the prescribed flight test hours have been completed and the aircraft is controllable throughout its range of speeds and throughout all maneuvers to be executed, has no hazardous operating characteristics or design features, and is safe for operation."

The Following Limitations Apply Outside of Flight Test Area:

1. Limitations 1, 3, 7, 8, 9, and 10 from Phase I are applicable.

2. This aircraft is approved for day VFR only, unless equipped for night VFR and/or IFR in accordance with FAR section 91.33 (new FAR section 91.205).

3. This aircraft shall contain the placards, markings, etc., required by FAR section 91.31 (new FAR section 91.9).

4. This aircraft is prohibited from acrobatic flight, unless such flights were satisfactorily accomplished and recorded in the aircraft logbook during the flight test period.

Figure 13-1

APPENDIX 9. SAMPLE LIST OF OPERATING LIMITATIONS (CONT'D)

5. No person may operate this aircraft for carrying persons or property for compensation or hire.

6. The person operating this aircraft shall advise each person carried of the experimental nature of this aircraft.

7. This aircraft shall not be operated for glider towing or parachute jumping operations, unless so equipped and authorized.

8. No person shall operate this aircraft unless within the preceding 12 calendar months it has had a condition inspection performed in accordance with FAR Part 43, appendix D, and has been found to be in a condition for safe operation. In addition, this inspection shall be recorded in accordance with limitation 10 listed below.

9. The builder of this aircraft, if certificated as a repairman, FAA certified mechanic holding an Airframe and Powerplant rating and/or appropriately rated repair stations may perform condition inspections in accordance with FAR Part 43, appendix D.

10. Condition inspections shall be recorded in the aircraft maintenance records showing the following or a similarly worded statement:

"I certify that this aircraft has been inspected on (insert date) in accordance with the scope and detail of appendix D of Part 43 and found to be in a condition for safe operation."

The entry will include the aircraft total time-in-service, the name, signature, and certificate type and number of the person performing the inspection.

____
Aviation Safety Inspector

____
Date Issued

____
Office Designation

TYPE OR PRINT ALL ENTRIES IN INK Form Approved OMB No. 2120-0022

US Department of Transportation
Federal Aviation Administration

AIRMAN CERTIFICATE AND/OR RATING APPLICATION

☐ MECHANIC
☐ AIRFRAME
☐ POWERPLANT

☐ REPAIRMAN

(Specify Rating)

☐ PARACHUTE RIGGER
☐ SENIOR ☐ MASTER
☐ SEAT ☐ CHEST
☐ BACK ☐ LAP

APPLICATION FOR: ☐ ORIGINAL ISSUANCE ☐ ADDED RATING

I. APPLICANT INFORMATION

A. NAME _(First, Middle, Last)_

B. SOCIAL SECURITY NO | C. DOB _(Mo., Day, Yr.)_ | D. HEIGHT IN | E. WEIGHT

F. HAIR | G. EYES | H. SEX | I. NATIONALITY _(Citizenship)_

J. PLACE OF BIRTH

K. PERMANENT MAILING ADDRESS
NUMBER AND STREET, P.O. BOX, ETC.
CITY
STATE ZIP CODE

L. HAVE YOU EVER HAD AN AIRMAN CERTIFICATE SUSPENDED OR REVOKED?
☐ NO
☐ YES _(If "Yes," explain on an attached sheet keying to appropriate item number)_

M. DO YOU NOW OR HAVE YOU EVER HELD AN FAA AIRMAN CERTIFICATE? ☐ NO ☐ YES
SPECIFY TYPE

N. HAVE YOU EVER BEEN CONVICTED FOR VIOLATION OF ANY FEDERAL OR STATE STATUTES PERTAINING TO NARCOTIC DRUGS, MARIJUANA, AND DEPRESSANT OR STIMULANT DRUGS OR SUBSTANCES? . . . . . . ☐ NO ☐ YES → DATE OF FINAL CONVICTION

II. CERTIFICATE OR RATING APPLIED FOR ON BASIS OF —

☐ A. CIVIL EXPERIENCE
☐ B. MILITARY EXPERIENCE
☐ C. LETTER OF RECOMMENDATION FOR REPAIRMAN _(Attach copy)_

☐ D. GRADUATE OF APPROVED COURSE
(1) NAME AND LOCATION OF SCHOOL
(2) SCHOOL NO | (3) CURRICULUM FROM WHICH GRADUATED | (4) DATE

☐ E. STUDENT HAS MADE SATISFACTORY PROGRESS AND IS RECOMMENDED TO TAKE THE ORAL, PRACTICAL TEST (FAR 65.80)
(1) SCHOOL NAME NO | (2) SCHOOL OFFICIAL'S SIGNATURE

☐ F. SPECIAL AUTHORIZATION TO TAKE MECHANIC'S ORAL PRACTICAL TEST (FAR 65.80)
(1) DATE AUTH. | (2) DATE AUTH. EXPIRES | (3) FAA INSPECTOR SIGNATURE | (4) FAA DIST. OFC.

III. RECORD OF EXPERIENCE

A. MILITARY COMPETENCE OBTAINED IN → (1) SERVICE | (2) RANK OR PAY LEVEL | (3) MILITARY SPECIALTY CODE

B. APPLICANTS OTHER THAN FAA CERTIFICATED SCHOOL GRADUATES: LIST EXPERIENCE RELATING TO CERTIFICATE AND RATING APPLIED FOR _(Continue on separate sheet if more space is needed)_

DATES—MONTH AND YEAR		EMPLOYER AND LOCATION	TYPE WORK PERFORMED
FROM	TO		

C. PARACHUTE RIGGER APPLICANTS, INDICATE BY TYPE HOW MANY PARACHUTES PACKED →

SEAT	CHEST	BACK	LAP

FOR MASTER RATING ONLY — PACKED AS A — ☐ SENIOR RIGGER ☐ MILITARY RIGGER

IV. APPLICANT'S CERTIFICATION

I CERTIFY THAT THE STATEMENTS BY ME ON THIS APPLICATION ARE TRUE
A. SIGNATURE | B. DATE

V.

I FIND THIS APPLICANT MEETS THE EXPERIENCE REQUIREMENTS OF FAR 65 AND IS ELIGIBLE TO TAKE THE REQUIRED TESTS | DATE | INSPECTOR'S SIGNATURE | FAA DISTRICT OFFICE

FOR FAA USE ONLY

Emp	reg.	D.O.	seal	con	iss	Act	rev	TR	s.n.	Srch	Prte	RATING (1)	RATING (2)	RATING (3)	RATING (4)

LIMITATIONS

FAA Form 8610-2 (2-85) SUPERSEDES PREVIOUS EDITION

Form 8610-2

Repairman's Certification

The Repairman's Certificate is applied for using the application from 8610-2, available from the local FAA offices. You should ask for this when you apply for your final inspection on your aircraft. You should also be familiar with the Appendix D of FAR part 43. (Items included in the Annual Condition Inspection.)

The Repairman's Certificate is only available to those who have built 51% or more of the specific aircraft they are having inspected. It is not available to **buyers** or builders that have completed less than the major portion. If you have a project that has been built by several people and passed from hand to hand and you want to buy it, you need to get the logs from all the builders to prove that you are the major portion builder of the aircraft, or your aircraft will be put in Exhibition — not Amateur Built — then you cannot carry passengers or work on the airplane.

Every twelve calendar months a condition inspection is performed in accordance with Appendix D of FAR part 43. The repairman has to include the aircraft total time in service, the name, the signature and the certificate type number of the repairman or A & P, who does the examination.

A&P mechanics must do the Annual Condition Inspection for those who are non-builders who own an amateur-built aircraft. On those aircraft where the builder has a Repairman's Certificate, it is **recommended** that from time to time the Annual Condition Inspection of those aircraft be done by an A&P simply as a check on the builder/repairman's work. One legal representative recommends that every other Annual Condition Inspection for a builder holding a repairman's certificate be done by an A&P mechanic.

Above — Outstanding workmanship and mirror smooth finish are evident on this Lancair 235. Understandably the aircraft has won numerous trophies for builder John Bittner, Clovis, CA.

Left — With a span of 17 ft. 8 in., and an empty weight of 375 lbs., this sleek little composite single place design by noted race pilot Nick Jones made its first public appearance at the 1990 EAA Oshkosh convention. Power is a 90 hp 3-cylinder AMW engine.

Weight And Balance

14-1 GENERAL

The center of gravity of an airplane may be defined, for the purpose of balance computations, as an imaginary point about which the nose heavy (−) moments and tail heavy (+) moments are exactly equal in magnitude. If the aircraft were suspended therefrom, it would have no tendency to rotate in either direction (nose up or nose down). The weight of the aircraft can be assumed to be concentrated at its CG. Weight and balance should follow the example in Advisory Circular 43.13-1A, Chapter 13.

14-2 C.G. RANGE

The allowable variation of the CG location is called the CG range. Since CG limits constitute the range of movement that the airplane CG can have without making the airplane unstable or unsafe to fly, the CG of the loaded airplane must be within these limits at takeoff, in the air, and on landing.

14-3 DATUM

The datum line, or simply datum, is a place or line in a plane used as a reference in order to show relative locations of objects (see diagram). In computations for weight and balance, the reference datum is vertical, and in the case below, we have located it at the front face of the propeller for convenience. All clockwise moments are positive and counterclockwise moments are negative. When the datum is located forward of the nose of the airplane, all moments must be positive. If the airplane were suspended at its nose, it is obvious that the tail would be downward; hence all movements are tail heavy, or positive.

It is very important to determine the exact location of the datum line before starting to solve the weight and balance problem.

14-4 MOMENT ARM

The moment arm, or merely the arm, is the horizontal distance in inches to the CG of an item measured aft of the datum. With the datum at the nose of the airplane, all measurements are aft and therefore positive.

The moment is the product of the weight and the arm. Thus, the moment of an item about the datum is obtained by multiplying the weight of the item by its horizontal distance from the datum. Moment is manifest as a tendency to cause rotation of the aircraft about its CG.

Before proceeding with the method of computing weight and balance, a thorough understanding of these terms is necessary.

14-5 EMPTY WEIGHT

The empty weight of an aircraft includes the weight of the airplane, powerplant, equipment which has a fixed location and is normally carried in the airplane, fixed ballast, and any other parts or equipment which are required during flight and are installed in the airplane. Fuel and oil are drained and only residual fuel and oil is included in the empty weight.

14-6 MAXIMUM WEIGHT

Maximum weight is the maximum authorized take-off weight of the aircraft and its contents. The weight and balance document included with any amateur built aircraft at its final inspection is a very important document. It is absolutely essential that this weight and balance document reflect the maximum gross weight the airplane will ever by flown at during its life. It is better to overstate the maximum gross weight rather than understate it.

14-7 USEFUL LOAD

Useful load is the empty weight subtracted from the maximum authorized takeoff weight for the aircraft. This load includes pilot, maximum oil, maximum fuel and baggage.

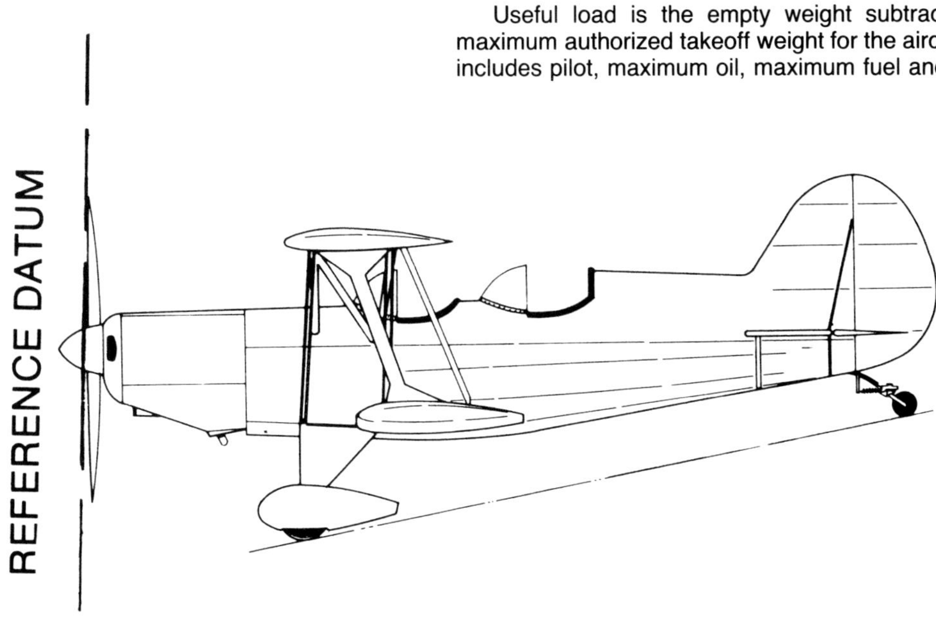

The Acro Sport undergoing final assembly prior to accomplishment of weight and balance on the aircraft.

14-8 WEIGHT CHECK

A weight check consists of checking the sum of the weights of all items of useful load against the allowable useful load of the aircraft.

14-9 EMPTY-WEIGHT CG

The empty-weight CG is the center of gravity of an aircraft in its empty condition.

14-10 OPERATING CG

The operating CG is the center of gravity of an aircraft in its loaded or operating condition.

14-11 WEIGHING POINT

The weighing points of an airplane are the points by which the airplane is supported at the time it is weighed. In this case, the landing gear and tail wheel are the weighing points.

14-12 TARE

Tare is the weight of the equipment necessary for weighing the airplane (such as chocks, blocks, etc.) which is included in the scale readings but is not a part of the actual weight of the airplane. Tare must be subtracted from the scale readings in order to obtain the actual weight of the airplane.

14-13 LEVELING MEANS

Leveling means are the reference points used for leveling the aircraft. In this case it is the top of the top fuselage longerons in the cockpit.

14-14 DETERMINATION OF EMPTY WEIGHT CENTER OF GRAVITY LOCATION

Never submit a weight and balance for an amateur-built aircraft that is not at the minimum and maximum that the aircraft will be flown. In event of a mishap the weight and balance you submitted will be checked and if you are out of CG you will be fined.

In order to obtain an accurate reading of the aircraft weight, the following procedures should be followed:

1. The aircraft should be weighed inside a closed building to avoid errors which may be caused by wind.
2. The airplane must be leveled longitudinally and laterally.
3. The accuracy of the scales must be established. Amateurs can use bathroom scales, provided they all have equal values. Additionally, heavier aircraft may use 2 scales, with a board across them for the wheels. The board, or tare weight, is subtracted in the calculations.
4. All items of equipment to be installed in the aircraft

and included in the empty weight should be in place for weighing.

5. The oil system should be drained. It is permissible to weigh the airplane with full oil, but the weight of the oil must be subtracted from the total weight to obtain the empty weight of the aircraft. The moment of the oil must be taken into consideration when the CG location is computed.

6. The fuel tank should be empty, or the fuel in the tank will have to be considered during CG location computation.

7. The weight of the tare should be recorded, either before or after weighing the airplane, and the weight should then be subtracted from the total weight as obtained from the scales.

8. When the airplane is in the level position, the exact horizontal distance between the supporting points must be recorded.

9. The weights of the right wheel, left wheel and tailwheel must be recorded to provide information needed for the CG determination.

The fundamental rule for determining the location for the CG for an airplane is: DIVIDE THE TOTAL MOMENT OF AN AIRPLANE TAKEN FROM THE DATUM LINE, BY THE TOTAL WEIGHT OF THE AIRPLANE. THE RESULT WILL BE THE DISTANCE OF THE CENTER OF GRAVITY FROM THE REFERENCE POINT.

In the figure, the Acro Sport II is weighed and it is found that the right wheel weight is 420 lbs., the left wheel weight is 410 lbs., and the tailwheel weight is 45 lbs.

These weights, with the location of the weighing points, give us the information we need to find the location of the CG. The weight of the blocks that are used in the scales must be removed as "tare".

We have established the datum line at 30" forward of the front face of the propeller hub. All moments will be determined from here. The moments and CG locations are found as follows:

WEIGHT AND BALANCE FORM

ITEM	LBS. NET	INCHES ARM	IN. LBS. MOMENT
Left Wheel	410	83.375	34,184
Right Wheel	420	83.375	35,018
Auxiliary Wheel	45	237.625	10,693

Empty Weight 875 Total Moment 79,895 in lbs.

Empty C.G. = $\frac{\text{Total Moment}}{\text{Empty Weight}} = \frac{79,895}{875} = \underline{91.3}$ inches

Care must be taken to insure that the proper sign is applied to each quantity expressed in a weight and balance computation. All moments are positive (+) in this instance, since the datum is forward of the nose. The weight of an airplane is always positive. Also, the weight of any item installed in the airplane is positive. The weight of any item removed from the airplane is negative (−). According to the rules of algebra, the product of numbers with like signs is positive, the product of numbers with unlike signs is negative.

14-15 LOADING THE AIRPLANE

The Acro Sport has a CG range within which the CG must lie if the aircraft is to be operated safely. In order to determine whether the loaded CG falls within the approved limits, it is necessary that two computations be made; one for most forward loading, and one for most rearward loading.

To determine the conditions for loading, we must consider the oil quantity and arm, fuel quantity and arm, pilot and arm, passenger and arm and baggage and arm. The fuel and oil weights tend to move the CG forward, the pilot, passenger and baggage weights tend to move the CG aft on the Acro Sport II. Our computations must include maximum quantities of these items.

Assuming that the empty weight of the airplane is 875 lbs., and the empty weight CG is 91.3 inches, we can load the airplane to determine whether the forward CG is within limits with maximum forward loading.

Unit weights have been established for weight and balance purposes. Gasoline is taken as 6 lb. per U.S. gallon, lubricating oil is 7.5 lbs. per U.S. gallon, 170 lbs. is allowed for the pilot, and 190 lbs. per aerobatic pilots (which includes parachute). Of course, exceptional pilot weight should be taken into consideration.

FORWARD C.G. EXTREME

ITEM	WEIGHT	ARM	MOMENT
Aircraft EW	875	91.3	79,895
Oil	15.2	52.3425	795.6
Pilot	120	148.5	17,820 (very lt. wt.)
Passenger	0	—	0
Fuel (20 gal.)	120	78.5	9,420
Baggage	0	—	0
TOTALS	1,130.2	—	107,931

Forward C.G. = $\frac{\text{Moment}}{\text{Weight}} = \frac{107,931}{1,130.2} = \underline{95.497}$ inches

The maximum forward position of the C.G. is at station 95.336, or 14% MAC. The above falls within the allowable C.G. range, since it falls aft of the most forward C.G. limit.

To check the airplane for rearward C.G. limit, we must use a maximum of all weights to the rear of the C.G. limit, and minimum of all weights forward of the rearward C.G. limit. Now the problem is arranged as follows:

REARWARD C.G. EXTREME

ITEM	WEIGHT	ARM	MOMENT
Aircraft EW	875	91.3	79,895
Oil	15.2	52.3425	795.6
Pilot	220	148.5	32,670 (very heavy)
Passenger	220	113.5	24,970 (very heavy)
Fuel (20 gal.)	6	78.5	471
Baggage	30	157.375	4,721.25
TOTALS	1,366.2	—	143,522.85

Rearward C.G. = $\frac{\text{Moment}}{\text{Weight}} = \frac{143,523}{1,366} = \underline{105.053}$ inches

Maximum aft C.G. limit is 105.226 inches. (37% MAC). This falls within the allowable C.G. range, since it falls forward of the most rearward C.G. limit.

It must be pointed out that the figures used in these examples are for the Acro Sport II prototype. The weights and arms will vary for each individual Acro Spot II, but the forwardmost and rearwardmost C.G. limits given here are applicable to all Acro Sport II's.

In the event that the C.G. doesn't fall within the approved range, it is necessary to correct the condition with ballast. Please see the following article for information on balancing an aircraft with ballast.

During the lifetime of the airplane, it may be desirable to change the type of equipment which is installed. In every case of such a change, it is necessary to figure the effect on weight and balance.

14-16 BALLAST

Any prejudice the builder may have against the use of ballast to balance the airplane should be gotten rid of at this point. On a prototype airplane, the use of ballast to achieve balance is normal and ordinary. Many plans sold to homebuilders have only sketchy information on C.G. locations and planes built to these plans may require ballast. Minor differences between builders may cause any airplane to need ballast. The same is true when changes are made in the engines or equipment installed. The only requirement is that the ballast be positively attached to structure strong and rigid enough to carry it.

Remember, unless you want to rebuild the horizontal tail, the C.G. location will absolutely determine the longitudinal stability and handling characteristics of your airplane.

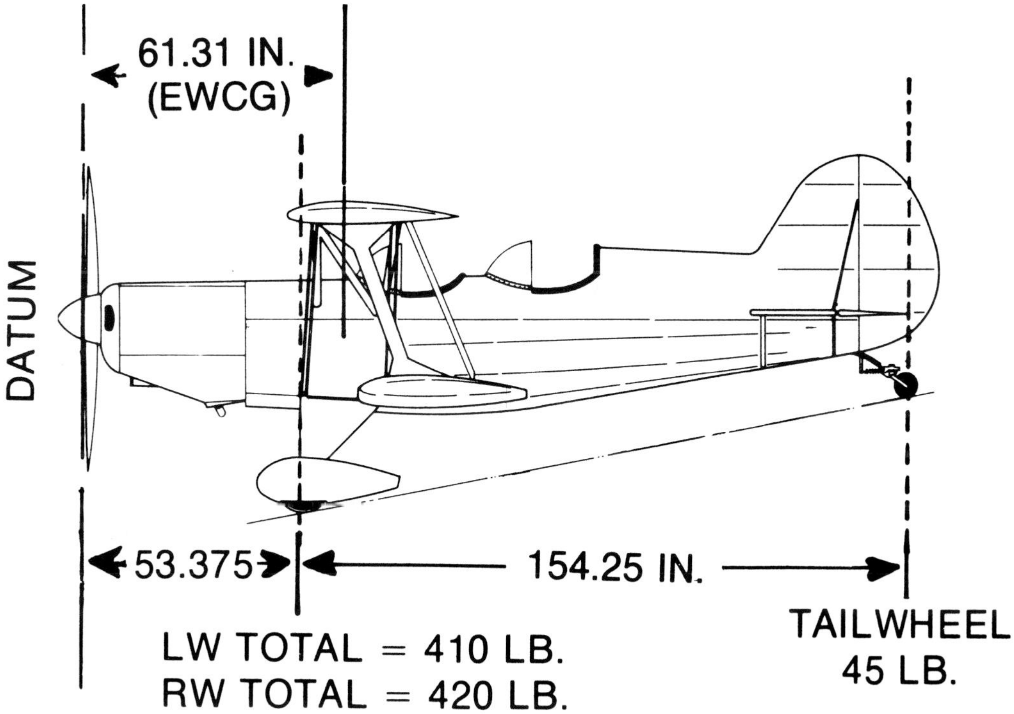

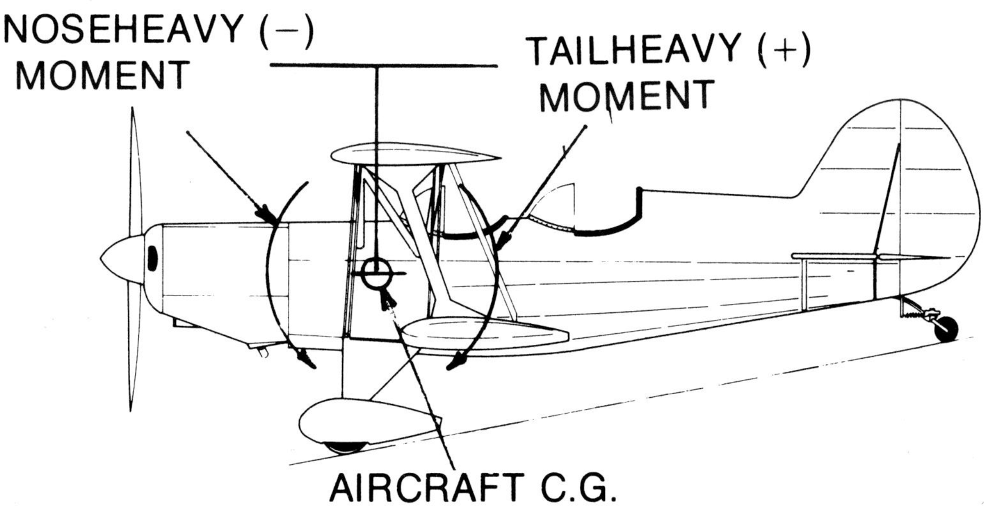

WEIGHT AND BALANCE INTRODUCTORY DRAWING

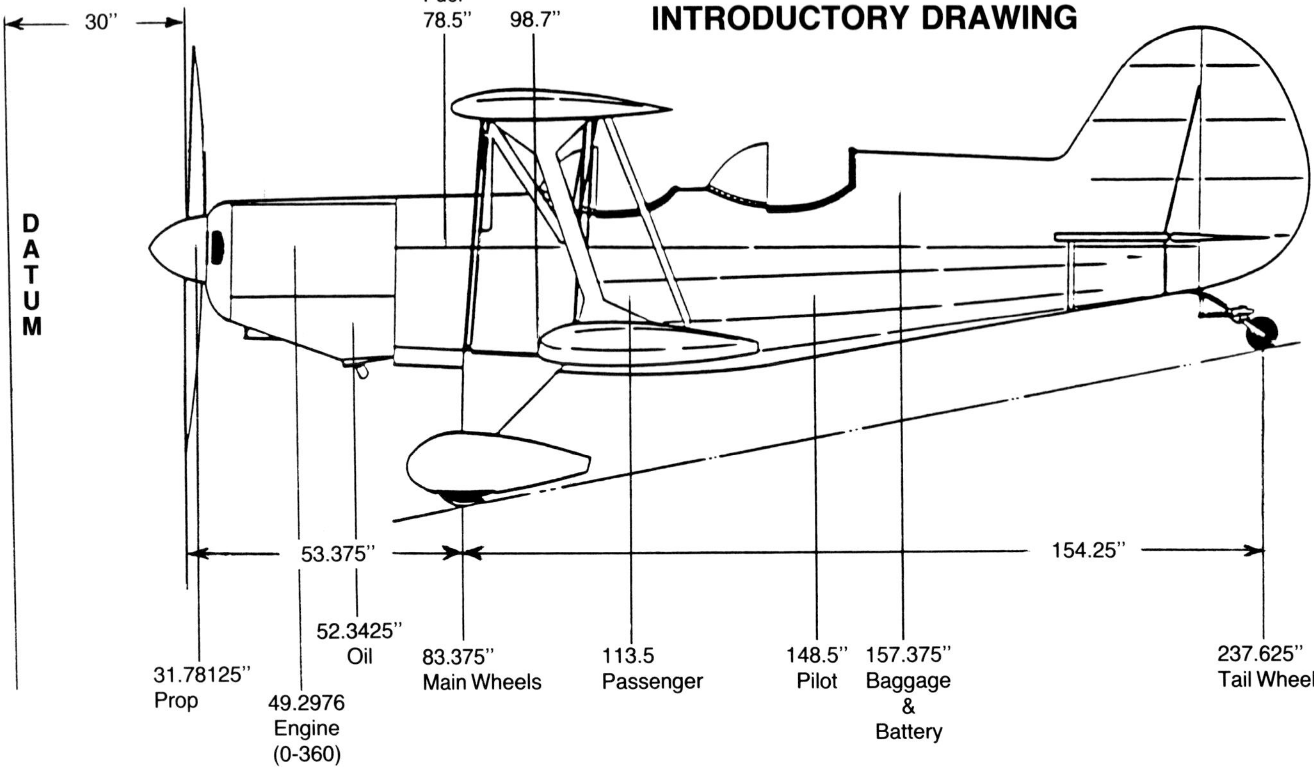

Practical Aircraft Balance

By S.E. "Ben" Owen - EAA #19702

14-17

The previous article dealt with calculating the weight and balance with examples from the Acro Sport II. This article will also give examples from the Acro Sport II, which are applicable to any other biplane or monoplane. The sketch will describe how to obtain the mean aerodynamic chord for a biplane. Determining the mean aerodynamic chord of a biplane can be solved by using books such as "Airplane Design Manual" by Frederick K. Teachmann or the book by Clemin, "Airplane Stress Analysis". These books are available through sources that sell older aircraft books.

The Acro Sport has a wing area of 152 sq. ft., horizontal stabilizer area of approximately 20 sq. ft. (13% of wing area) and a vertical stabilizer and rudder area of 13.8 sq. ft. approximately (9% of wing area). This, combined with the excellent stable characteristics of the Munk M-6 airfoil, have given a quite satisfactory C.G. range.

14-18

The Mean aerodynamic chord determination is very important to the design of any airplane, and is a starting point for all design calculations. In our choice of a datum line, for instance, we moved the datum 30" in front of the front face of the propeller flange, so that all moments would be positive when designing for a very light engine weight. This eliminates a necessity for use of negative moments.

In our method here, we have removed the engine, cowling, propeller (all firewall forward) from the aircraft and replaced it with the same components of both lighter and heavier engines. In addition, the Acro Sport II was not origi-

BIPLANE MEAN AERODYNAMIC CHORD

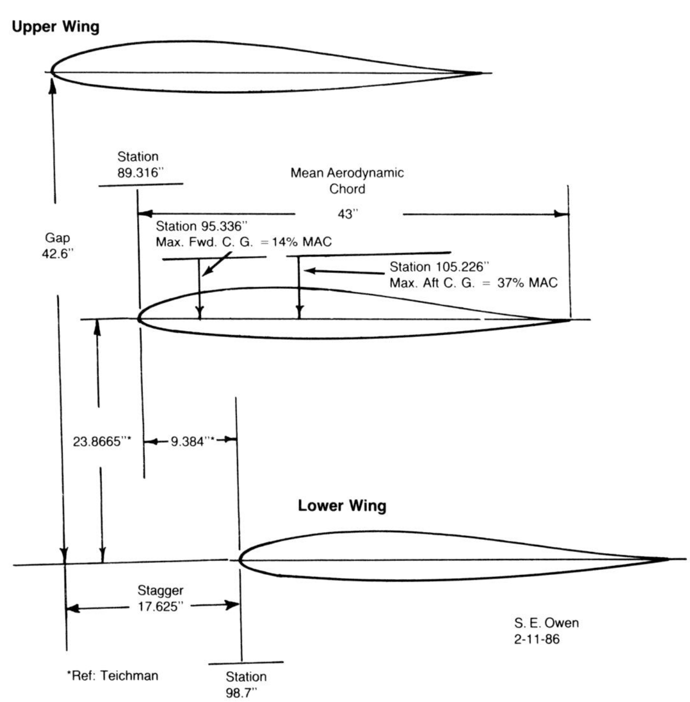

nally designed for an electrical system. However, many people might like to add these to either their Acro Sports or to their other amateur built aircraft. For this reason, the information on this is also included. The following are the reference weights for our calculation.

14-19

A List	**Engine baffles**	**3 lbs., 5 oz.**
	Nose bowl	**4 lbs.**
	Cowling sheet metal, etc.	**9 lbs., 8 oz.**
	Carburetor heat box	**3 lbs.**
	Fuel and Oil lines	**3 lbs.**
	Oil	**7 lbs., 8 oz.**
	Engine mount and bolts	**8 lbs., 3 oz.**
	TOTAL	**38.5 lbs.**

Add 1 or 2 lbs. more for a longer cowl. Weight of the A list will be centered on the engine C.G.

B List	**Sensenich metal propeller 76/**	**42 lbs., 7 oz.**
	Propeller bolts and nuts	**1 lb., 6 oz.**
	Spinner	**2 lbs., 11 oz.**
	TOTAL	**46.5 lbs.**

Arm of the B list is at the propeller center line. For wooden propellers, subtract about 20 lbs.

C List	**Battery**	**21 lbs.**
	Battery box	**1.5 lbs.**
	TOTAL	**22.5 lbs.**

Arm is aft of the rear seat of an Acro Sport II.

D List	**Over voltage relay**	**10 oz.**
	Voltage regulator	**1 lb., 3 oz.**
	Wiring	**3 lbs.**
	Starter switch	**1 lb., 1 oz**
	Anmeter	**6 oz.**
	Miscellaneous panels, breakers switches and wires	**6 lbs., 4 oz.**
	TOTAL	**13 lbs.**

Arm in the Acro Sport II at the pilots instrument panel.

The following calculations are to determine the center of gravity of all firewall forward items and electrical system components. An important number in these calculations is the distance that the firewall forward center of gravity is forward of the engine center of gravity. We will use the firewall forward center of gravity location in some of our calculations, but the **difference** between the firewall forward C.G. and the engine C.G. is primarily what we are looking for here. This will be used later to determine engine mount length. I also will introduce negative weights and moments to familiarize you with the procedure used.

14-20 - SECTION A

THE CENTER OF GRAVITY - LYCOMING 0-360 ENGINE WITHOUT STARTER AND GENERATOR OR ALTERNATOR

ITEM	WEIGHT IN LBS.	ARM IN INCHES	MOMENT IN INCH/LBS.	REMARKS
Engine	288	14.5622	4,193.9136	Engine is 2 lbs. over "Normal" weight
Starter	- 18	4 aprox	- 72	
Alternator	- 13	6 aprox	- 78	
TOTALS	257		4,043.9136	

Please note the datum in the above is the front face of the rear propeller flange.

arm = moment / weight = 4043.9136 in./lbs. / 257 lbs. = 15.7351 in.

14-20 - SECTION B

DETERMINING THE CENTER OF GRAVITY OF THE 0-360A WITHOUT STARTER AND ALTERNATOR BUT INCLUDING THE PROPELLER AND COWL OR WHAT IS KNOWN AS "FIREWALL FORWARD"

ITEM	WEIGHT IN LBS.		ARM IN INCHES	MOMENT IN INCH/LBS.
Engine	257			
"A" List-Cowl	38.5			
		295.5	15.7351	4,649.722
"B" List-Prop		46.5	- 1.78125	- 82.8281
TOTAL		342.		4,566.8939

Note: Prop is 3.5625" deep. 3.5625" divided by 2 = - 1.78125" for propeller arm. Please note the datum is again at the front face of the rear prop flange.

arm = moment / weight = 4,566.8939 in./lbs. / 342 lbs. = 13.3535 inches.

15.7351
-13.3535
2.3996" = amount the firewall forward C.G. is ahead of the engine alone C.G.

Since we have been using the common datum for engine installation drawings of the front face of the propeller plan, we'll have to transpose this to our new center of gravity:

14-20 - SECTION C

Old Datum	**13.3535**
Prop thickness	**3.5625**
New Datum location	**30.**
46.9160	**from new datum to firewall forward C.G.**

DATUM	WEIGHT	ARM	MOMENT
Old: Front face, rear Prop Flange	342	13.3535	4,566.8939
New: 30" forward of Prop Front Face	342	46.9160	16,045.272

14-20 - SECTION D

PROTOTYPE ACRO SPORT II WEIGHTS

The next procedure is to determine the weight of the airframe only, without the firewall forward weights.

ITEM	WEIGHT IN LBS.	ARM IN INCHES	MOMENT IN INCH/LBS.	REMARKS
Acro Sport II Empty Weight	875	91.3	79,895	See previous article
-0-360, Firewall Forward	-342	46.9160	-16,045.272	
TOTAL	533		63,849.728	

arm = moment / weight = 63,849.728 in./lbs. / 533 lbs. = 119.79311 inches

The center of gravity Acro Sport II airframe only - firewall forward items = 533 lbs. and 119.79311 inches.

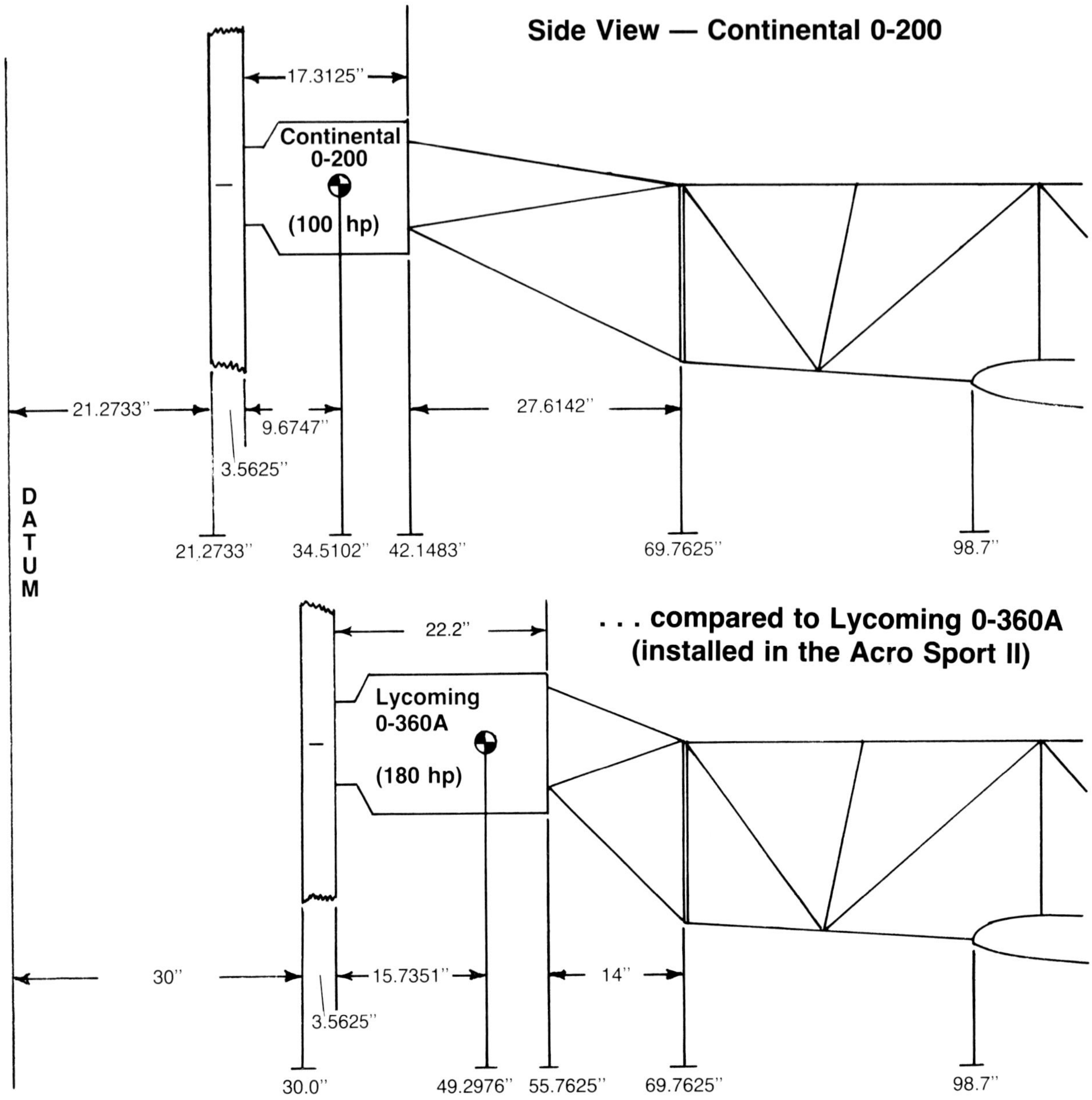

14-21 INSTALLING A LIGHTER ENGINE

A. The 0-200 weight with starter and generator equals 212.55 lbs. Its C.G. is 9.6747 inches back of the propeller flange front face. The following calculations subtract the starter and generator from the overall engine weight.

ITEM	WEIGHT IN LBS.	ARM IN INCHES	MOMENT IN INCH/LBS.	REMARKS
Engine	212.55	9.6747	2,056.3574	C.G. from Teledyne
Starter	- 24.	23.5 aprox	-564	Weight from Teledyne
Generator	- 15.	23. aprox	-345	Measured Distances
TOTAL	173.55		1,147.3574	

Note: The datum was the front face of the rear prop flange.

arm = $\frac{\text{moment}}{\text{weight}}$ = $\frac{1{,}147.3574}{173.55 \text{ lbs.}}$ = 6.6111 inches = C.G. engine alone

DETERMINE THE 0-200 FIREWALL FORWARD WEIGHT (NO STARTER, NO GENERATOR-ALTERNATOR)

ITEM	WEIGHT IN LBS.		ARM IN INCHES	MOMENT IN INCH/LBS.	REMARKS
Engine	173.55				
"A" List-Cowl	40.5				Add 2 lbs.
		214.05	6.6111	1,415.0845	long
"B" List		46.5	-1.78125	- 82.8281	cowl
TOTAL		260.55		1,332.2564	

Note: Datum front face, rear propeller flange.

arm = $\frac{\text{moment}}{\text{weight}}$ = $\frac{1{,}332.2564}{260.55 \text{ lbs.}}$ = 5.1132" or C.G. location of firewall forward items

Engine & Cowl C.G.	6.6111
Firewall Forward C.G.	-5.1132
	1.4979" = amt. engine C.G. is aft of Firewall Fwd C.G.

14-21 - SECTION C

INSTALLING THE 0-200

ITEM	WEIGHT IN LBS.	ARM IN INCHES	MOMENT IN INCH/LBS.	REMARKS
Airframe ASII				
Empty Weight	533	119.7931	63,849.728	See 14-20,
0-200 F.F.	260.55	(d) 33.0124	(c) 8,601.387	Section d
TOTAL	(a)793.55	91.3	(b)72,451.115	Minus Starter & Gen.

SOLVE AS FOLLOWS:

a. Solve for weight (add).
b. Solve for total moment - M = WxA = 793.55 x 91.3 = 72,451.115 (Note: Use 91.3 inches as the arm from the previous weight and and balance.)
c. Solve for Engine Moment = Total moment = 72,451.115
   -Airframe moment = 63,849.728
   Engine moment = 8,601.387
d. Solve for Firewall Forward (0-200) arm
   arm = moment / weight = 8,601.387 / 260.55 lbs. = 33.0124
e. Solve for engine center of gravity.
   Firewall Forward C.G. = 33.0124
   Difference 1.4978
   TOTAL is Engine C.G. = 34.5102

14-22 - SECTION A

ADDING AN ELECTRICAL SYSTEM

First, you must determine the new weight of the airframe alone, including the battery and electrical system. For the larger engines, the battery is located aft of the pilot's seat. We are estimating the total electrical system wiring, etc., to be at the pilot's instrument panel as far as the lever arm goes. The following are the calculations:

ITEM	WEIGHT IN LBS.	ARM IN INCHES	MOMENT IN INCH/LBS.	REMARKS
AS II				
Airframe	533	119.7931	63,849.728	
Battery	22.5	157.7625	3,549.6562	Aft of Pilot's seat
Electric	11	130.8675	1,439.5425	On station 4,
TOTAL	566.5		68,838.926	top longeron

arm = moment / weight = 68,838.926 / 566.5 lbs. = 121.5162 inches at 566.5 lbs.

14-22 SECTION B

DETERMINE THE FIREWALL FORWARD CENTER OF GRAVITY FOR THE 0-200 WITH STARTER AND ALTERNATOR

Note: The datum again is the front face of the rear propeller flange for this calculation.

ITEM	WEIGHT IN LBS.		ARM IN INCHES	MOMENT IN INCH/LBS.	REMARKS
0-200	212.55				See 14-21
"A" List	40.5				Long cowl
		253.05	9.6747	2,448.1828	
"B" List		46.5	-1.78125	- 82.8281	
TOTAL		299.55		2,365.3547	

arm = moment / weight = 2,365.3547 / 299.55 lbs. = 7.8964

Difference: Engine C.G. 9.6747
Firewall Forward C.G. -7.8964
1.7783

14-22 SECTION C

Use Formula as in 14-22(b) and obtain same data for other engines: (weight includes starter and alternator or generator). Datum is still the front face of the rear propeller flange.

TYPE	ENGINE LENGTH	ENGINE WT.	ENGINE CG	FRWL FWD. WT.	FRWL FWD. CG	DIFFERENCE
0-200	17.3125	212.55 lbs.	9.6747	299.55 lbs.	7.8964 in.	1.7783 in.
0-235	22.2	240	14.75	326	12.3920	2.358
0-320	22.2	272	14.04	357	11.9792	2.0606
0-360A	22.2	286	14.14	371	12.1444	1.9955
10-360A	22.2	332	14.24	417	12.4535	1.7865

14-23 - SECTION A

DETERMINING ENGINE MOUNT LENGTH FOR THE 0-200 WITH ELECTRIC

ITEM	WEIGHT IN LBS.	ARM IN INCHES	MOMENT IN INCH/LBS.	REMARKS
A.S. II Airframe	566.5	121.5162	68,838.926	See 14-23A
0-200 Frwl.Fwd.	299.55	(d) 34.15603	(c)10,231.439	
TOTAL	(a)866.05	91.3	(b)79,070.365	

a. Add for empty weight = 866.05
b. Use 91.3 as final moment - see prior weight and balance moment.
   Moment = weight x arm / Moment = 866.05 x 91.3 = 72,070.365
c. Solve for engine moment 79,070.365
   -68,838.926
   10,231.439
d. Solve for 0-200 Firewall Forward arm
   Arm = moment / weight = 10.231.439 / 299.55 lbs. = 34.1560
e. Solve for Engine C.G. location. 34.1560
   1.7783
   35.9343 inches
f. Solve for engine mount length.

1). Engine length 17.3125
 - Eng. CG to Flange 9.6747
 Sub-Total 7.6378

2). Subtotal 7.6378
 + Eng. CG Loc. 35.9343
 Location Mt. 43.5721

3). Firewall 69.7625
 - Location of Mt. 43.5721
 Total Mt. Length 26.1904
 inches for 0-200 with electrical system.

14-23 - SECTION B

We can see that a better solution to the 0-200 would be to mount the battery and the battery box forward of the firewall to reduce the engine mount length. Due to heat, an air duct from the nose bowl to the battery box is recommended with this installation.

0-200 E MOVING THE BATTERY FORWARD

ITEM	WEIGHT	ARM	MOMENT
Acro Sport II			
Air Frame	533	119.7931	63,849.728
Battery	22.5	65.7625	1,479.6562
Electric	11.	113.5	1,248.5
TOTAL	566.5		66,577.884

arm = moment / weight = 66,577.884 / 566.5 lbs. = 117.5249 in. at 566.5 lbs.

ITEM	WEIGHT	ARM	MOMENT	REMARKS
Acro Sport II				
Airframe	566.5	117.5249	66,577.884	Electric Fwd.
0-200 Frwl Fwd.	299.55	(d) 41.7042	(c)12,492.481	Solve a,b,c,d
TOTAL	(a)866.05	91.3	79,070.365	as in previous problems.

The engine C.G. is 41.7042 and solve for the mount length:
17.3125 - 9.6747 =7.6378
The engine C.G. 41.7042 + 7.6378 = 49.342
69.7625 - 49.342 = 20.4205 for Mount Length, 0-200 with electrical system and battery box situated forward of the firewall.

14-23 - SECTION C

SOLVE FOR MOUNT LENGTHS FOR OTHER ENGINES USING 14-23(a) AS AN EXAMPLE

TYPE	EMPTY WT.	MOUNT LNGTH.	REMARKS
0-235	892.5	21.1621	Includes starter,
0-320	923.5	16.1898	alternator, generator,
0-360A	937.5	14.54575	etc. battery aft of
I0-360A	983.5	9.7651	the pilots seat.

14-24

If the builder would like to maintain the mount length of 14 inches for the I0-360A, the following can be done:

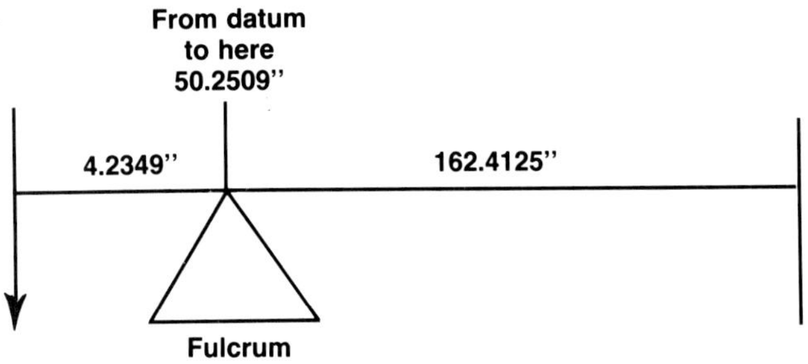

4.2349 x 417 = 1765.9533 in/lbs.

$$\frac{\textbf{1765.9533 in/lbs.}}{\textbf{162.4125"}} = \textbf{10.87 lbs.}$$

To move the mount length out 4.2349 inches so that it will be 14 inches long, this will be a weight of 417 lbs. times an arm of 4.2349 inches, for a moment of 1,765.9533 inch/lbs. Dividing this by 162.4125, we come up with a 10.87 lbs. This weight to be added at the tail spring is 10.87 lbs. and is a very simple "lever" drawing. The fulcrum is the location of the center of gravity of the firewall forward weight.

14-25 SECTION A

CORRECTING A NOSE HEAVY CONDITION USING THE ADVERSE MOMENT PRINCIPLE

The following process was developed by long time EAA'er, the late Noel Becar.

Step 1 - solve for adverse moment.

$$\frac{\textbf{Aircraft Empty Weight}}{\textbf{875 lbs.}} \times \frac{\textbf{Distance Between Existing And Desired C.G.}}{\textbf{2.61}} = \frac{\textbf{Adverse Moment}}{\textbf{2,284 lbs.}}$$

CORRECTING A NOSE HEAVY CONDITION (Empty Weight)

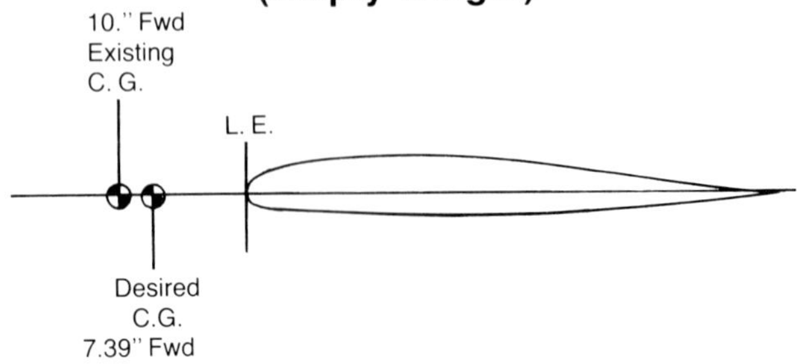

Step 2 - Compute distance to move engine back.

$$\frac{\textbf{Adverse Moment}}{\textbf{Engine weight}} = \frac{\textbf{2284}}{\textbf{357}} = \textbf{6.4"}$$

14-25 SECTION B

CORRECTING A TAIL HEAVY CONDITION

Step 1 - Solve for Adverse Moment.

$$\frac{\textbf{Aircraft Empty Weight}}{\textbf{875 lbs.}} \times \frac{\textbf{Distance Between Existing And Desired C.G.}}{\textbf{3.39 inches}} = \textbf{2966}$$

Step 2 - Compute distance engine must be moved forward.

$$\frac{\textbf{Adverse Moment}}{\textbf{Firewall Fwd Wt.}} = \frac{\textbf{2966}}{\textbf{357}} = \textbf{8.3 inches}$$

CORRECTING A TAIL HEAVY CONDITION (Empty Weight)

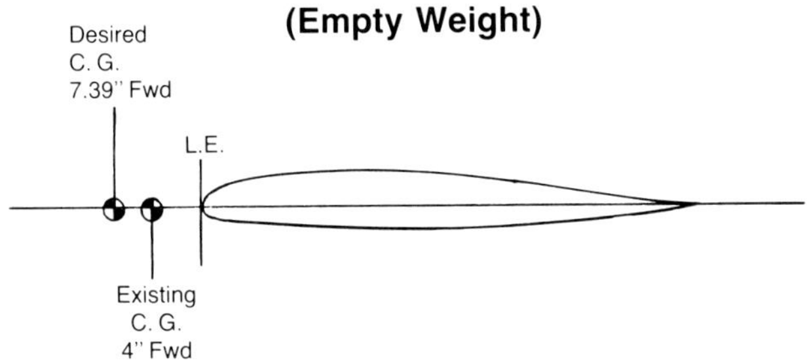

The Pober Super Ace is a new design by EAA founder Paul Poberezny. The prototype shown right, is powered with an 85 hp Continental C85-8 engine, but powerplants up to 150 hp can be installed. Aircraft is constructed of conventional 4130 steel tubing with wood ribs and spars in the wing, fabric covered. For open cockpit, fun flying that is economical, an aircraft like the Pober Super Ace is hard to beat.

PART II

General Design, Construction And Fabrication

15-1 GENERAL

The methods of fabrication used to construct the primary structure of custom-built sport aircraft should be such as to produce a uniformly sound structure. This structure should also be reliable with respect to maintenance of the original strength under reasonable service conditons.

15-2 MATERIALS AND WORKMANSHIP

The primary structure should be made from materials which experience or conclusive tests have proved to be uniform in quality and strength and to be otherwise suitable for the construction of the homebuilt.

The suitability of all design details should be firmly established prior to construction. The primary structure should not incorporate design details or materials which experience has shown to be unreliable or unsatisfactory.

Hardware such as bolts, pins, screws, nuts, wires, terminals, etc., used in the primary structure should be of aircraft standards as established by the SAE or as established by government aircraft standards and specification.

Self locking nuts may not be used on any bolt subject to rotation in operation.

Acceptable methods, techniques and practices considered universally acceptable in aircraft construction are detailed in FAA Advisory Circulars AC-43.13-1 and AC-43.13-2 (See Section 5), and builders are encouraged to obtain these two technical manuals for their guidance. Although the applicable methods outlined in the manuals and herein described are presented as compliance suggestions and as accepted practices for the construction and fabrication of amateur-built aircraft, the material presented should not be construed as the only satisfactory practice in the achievement of the fabrication objectives.

Workmanship of the primary structure should be of sufficiently high grade as to insure proper continued functioning of all parts. Each piece and component should be fabricated within the tolerances established for that part and should be properly finished and free from blemishes, cracks and scratches.

The methods of fabrication used must produce consistently sound structures. If a fabrication process such as gluing, spot welding, welding, or heat-treating requires close control to reach this objective, test samples should be made to verify results.

Material strength properties must be based on enough tests of material to establish adequate structural integrity.

The structural materials must be chosen so that the probability of any structure being understrength because of material variations is extremely remote. The builder is cautioned against the unwarranted and indiscriminate "beefing up" of the structure as it is a self defeating process.

Means must be provided to allow inspection, including the inspection of principal structural elements and control systems, close examination, repair, and replacement of each part requiring maintenance, adjustments for proper alignment and function, lubrication or servicing.

All bolt holes should be true and meet acceptable tolerances for the fit needed.

Bolts should be installed with the heads up or forward into the slipstream for safety. Acceptable tolerances and standards with regard to the bearing surfaces or areas and the thread clearances should be met.

15-3 PROTECTION OF STRUCTURE

All members of the structure should be suitably protected against deterioration or loss of strength in service due to weathering, corrosion, embrittlement, abrasions, and other causes.

Adequate provision for ventilation and drainage of all parts of the structure will contribute considerably to the protection of the structure from the effects of altitude and the environment.

15-4 STATIC OR MASS BALANCE

In order to prevent serious flutter characteristics during flight operation, the control surfaces (aileron, elevator and rudder) of most high performance homebuilts may require mass balance. This is not to be construed that the slower and lighter homebuilts are safe from flutter . . . or exempt from the need for static balance of the control surfaces in some cases.

15-5 FUSELAGE DESIGN CRITERIA

The fuselage is designed to carry the pilot and passenger(s), support the tail surfaces and landing gear with as little weight and drag as is consistent with the general purpose of the conventional aircraft.

The nose of the fuselage should be made sufficiently strong to give the pilot reasonable protection in case of a crash.

As far as practicable, the nose structure ahead of the cockpit should be designed to progressively collapse from the extreme nose towards the cockpit, upon head-on impact. This means that structure should be progressively stronger from the nose towards the rear. Such shock absorption minimizes the effects of sudden deceleration encountered in a head-on impact.

15-6 COCKPITS AND CABINS

Relatively rigid structural members or rigid mounted items of equipment which might be struck by the head, arm, knee, et cetera, should be padded. Padding preferably should be of foam polystryrene or unicellular polyvinal chloride material

Left — Fuselage sides tack welded and positioned to form the basic box structure. Important at this stage to have everything in perfect alignment — so take care that all measurements are accurate. Mistakes here can complicate later construction and actually affect the aircraft's performance. Note wood blocks on work table to hold tubing in place, and cross wires to square up the sides, top and bottom.

Below — Welding for many builders, is very enjoyable. Actually a way to relax and unwind. Some EAA'ers have actually developed the true artists' touch — their welds having the look and strength of perfection.

Simple construction of Pixie wing showing built up ribs.

Constant chord wings have many ribs that are identical in size and shape. A simple jig, as shown above, makes duplicating these ribs a fast, simple matter. Can be used even when making the compression ribs.

rather than ordinary foam or sponge rubber, which are ineffective.

Instrument panels should be smooth with top edges curved in a substantial radius. It is preferred that openings for instruments should have beveled instead of sharp edges.

Control wheels and control sticks should be free of sharp edges, hooks or projections which might cause injury in a minor accident.

15-7 SEATS

Seats should be adequate to withstand downward loads without failure of the seat or injury to the pilot. They should be securely fastened in place, whether or not the safety belt load is transmitted through the seat.

Consideration should be given in the seat installation to the fact that under some types of operations parachutes may be desired or required. Provisions, therefore, should be made for parachutes.

A shoulder harness is also considered necessary for all operations and should be installed, and properly attached to the fuselage primary structure.

15-8 CABIN VENTILATION AND VISIBILITY

The pilot's compartment should be free from glare and reflections that could interfere with the pilot's vision, and be so designed as to afford suitable ventilation both on the ground and under normal flying conditions.

15-9 TURNOVER STRUCTURE

The fuselage and cabins should be designed to protect the passengers and pilot in the event of a complete turnover. Adequate provision should be made to permit egress of the passengers and pilot in such an event.

It should be possible to open canopies and doors from the outside as well as from the inside. A clear fore and aft opening of not less than 24 inches is desirable. These recommendations apply also to the hinged, sliding, or removable canopy types of enclosures.

15-10 SAFETY EQUIPMENT

Safety belts should be so attached that no part of the anchorage is likely to fail at a load lower than that corresponding with the seat belt's capacity.

If the seat belt is attached to the seat, the structural integrity should be carried through to the primary structure.

The seat belt should have a metal-to-metal buckle and be of the quick release type.

The belt should be so installed as to pass across the two forward prominences of the pelvis, and should form at least a 45° angle with the horizontal axis of the aircraft. A belt so installed will pull downward as well as backward. The seat belt and shoulder harness and the hard-point should tolerate 40 G's.

The shoulder harness should be installed so that it passes backward over the shoulders at an angle approximately perpendicular to the spine. It should be attached to solid structural members and be capable of taking loads equal to that of the seat belt.

15-11 LANDING GEAR

To insure good ground handling characteristics it is essential that the landing gear be accurately aligned while the aircraft is in a level attitude.

Gear legs should be attached to the primary structure in such a manner as to avoid localized stress concentrations in landing.

While the installation of properly designed and constructed shock absorbers is highly recommended, a rigid gear with large low pressure tires may be successfully incorporated in lighter and slower aircraft.

The installation of brakes is recommended and either standard hydraulic brakes or any form of mechanical brake is acceptable. Brakes should operate in a completely normal manner. In the case of mechanical brakes, it is important that slack or play in the linkage be held to an absolute minimum to insure adequate braking.

15-12 RETRACTABLE LANDING GEAR

The landing gear and retracting mechanism, including the wheel well doors, should withstand flight loads, including loads resulting from all conditions encountered with the landing gear extended at any speed up to at least 1.6 times the stalling speed (Vs 1).

There must be positive means, other than the use of hydraulic pressure, to keep the landing gear extended.

A landplane without a manually operated landing gear system should have an auxiliary means of extending the gear in emergencies.

The proper functioning of the retracting mechanism should be established by operational ground tests while the aircraft is blocked up.

There must be a means to indicate to the pilot when the wheels are secured in their extreme positions.

For a landplane, an aural, or any equally effective landing gear warning device, should be provided.

15-13 HULLS AND FLOATS

The hull of a hull seaplane or amphibian of 1500 pounds or more maximum weight must have watertight compartments designed and arranged so that the hull, auxiliary floats, and tires (if used), will keep the aircraft afloat in fresh water.

Each main float should have a buoyancy of 80% in excess of that required to support the maximum weight of the seaplane or amphibian in fresh water. Additionally, enough watertight compartments to provide reasonable assurance that the seaplane or amphibian will stay afloat if any two compartments of the main floats are flooded.

The hulls of hull seaplanes or amphibians of less than 1500 pounds maximum weight need not be compartmented.

15-14 METAL COVERED SURFACES

Metal covered wings and other components should be free from buckling or wrinkling of the metal covering. Deflections or deformations at low load factors which may result in fatigue failures should also be avoided.

Metal fuselage on gear showing aluminum turtledeck, fin and stabilizer held in place by Clecos prior to riveting.

Landing Gear

Editor's Note: So often in the world of aeronautical design, very few people are exposed to the effort that goes into the actual designing and building of a certain part of an aircraft. For instance, the basic landing gear of any aircraft is subject to many loads and stresses during its service life.

The requirements for the proper design consistent with good aeronautical practices have been with us for many years! The following is taken from the U.S. Department of Commerce Aeronautics Bulletin No. 7-A entitled "Airworthiness Requirements for Aircraft".

SEC. 34. DESIGN CONDITIONS

(A) The following conditions represent the minimum amount of investigation required for conventional landing gear. For unconventional types it may be necessary to investigate other landing attitudes, depending on the arrangement and design of the landing gear members. Consideration will be given to a reduction of the specified load factors in cases where it can be proved that the shock-absorbing system will positively limit the acceleration factor to a definite lower value.

SEC. 35. LEVEL LANDING

(A) The minimum level landing applied load factor shall be obtained as follows:

(1) For airplanes of over 1,000 pounds gross weight —

$$\mathbf{n} = 2.80 + \frac{9.000}{W + 4,000}$$

where **W** = gross weight in pounds.

(2) For airplanes of less than 1,000 pounds gross weight —

$$\mathbf{n} = 3.00 + 0.133\ s.$$

where **s** = wing loading, in pounds per square foot.

(3) In any case the value of applied load factor **n** as determined from (1) or (2) need not exceed 4.33.

(B) The resultant of the ground reaction shall be assumed to be a force lying at the intersection of the plane of symmetry and a plane in which are located the axles and the center of gravity of the airplane less chassis. The propeller axis (or equivalent reference line) shall be assumed horizontal and the basic value of the vertical component of the resultant ground reaction shall be equal to the gross weight of the airplane minus chassis and wheels. The horizontal component shall be of the magnitude required to give the resultant force the specified direction, except that it need not be greater than 25 percent of the vertical component. When this exception applies, the resultant force will not pass through the center of gravity. Methods of accounting for the unbalanced moment are outlined in Aeronautics Bulletin No. 26, section 39 (C).

(C) If a sliding element instead of a rolling element is used for the landing gear, a horizontal component of one-half of the vertical component shall be used to represent the effect of ground friction, except that ski gear which is designed and used only for landing on snow and ice may be designed for the same horizontal component as wheel gear.

(D) The resultant of the ground reaction shall be assumed to be divided equally between wheels and to be applied at the axle at the center of the wheel. The shock-absorber unit and tires shall be assumed to be deflected to half their total travel, unless it is apparent that a more critical arrangement could exist.

SEC. 36. THREE-POINT LANDING

(A) The applied load factor shall be the same as for the level landing condition.

(B) The basic value of the sum of the ground reactions shall be the gross weight of the airplane less chassis. The total load shall be divided between the chassis and tail skid in inverse proportion to the distances, measured parallel to the ground line, from the center of gravity of the airplane less chassis to the points of contact with the ground. The load on the chassis shall be divided equally between wheels. Load shall be assumed to be perpendicular to the ground line in the 3-point landing attitude, with all shock absorbers and tires deflected to the same degree as in the level landing condition. The tail wheel or skid installation shall also be investigated for this condition.

Left — Landing gears such as this retractable gear take many hard loads, including the side loads of high-speed turnoff at maximum gross weights. Short travel of oleo strut give minimum travel to absorb shock loads of hard landing or rough fields. Right — Welded tripod type gear will absorb shocks only in the deflection of tires — there is no other shock absorbing mechanism provided. Very gentle landings are called for.

Left — Cub type gear installed on the EAA Biplane. External shock cords are usually covered with aluminum cuffs, as shown here. Cuffs can also be fabricated of canvas, leather or fireproof synthetic material. Right — Designed so that shock cords are inside the fuselage this gear looks cleaner and it will also be aerodynamically cleaner adding several miles to cruise speed.

SEC. 37. SIDE LOAD AND ONE-WHEEL LANDING CONDITIONS

(A) **Side load** — The landing gear and adjacent portions of the fuselage and cabane shall be analyzed for the following loading:

(1) An **applied** load equal to the gross weight of the airplane (less chassis) times one-tenth of the level landing applied load factor shall be assumed to act inward at the point of contact of one wheel with the ground in a direction normal to the plane symmetry.

(2) The load specified in paragraph (1) shall be balanced by inertia loads, the resultant of which passes through the center of gravity of the airplane less chassis. These loads can be distributed between panel points adjacent to the landing gear.

(3) The unbalanced moment of the side load about the center of gravity will be resisted by angular inertia loads. It will be satisfactory to apply the necessary couples at the wing attachment points, assuming that the unbalanced turning moment is resisted entirely by the angular inertia of the wing structure.

(4) The shock absorbers and tires shall be assumed to be deflected to the same degree as in level landing.

(5) No vertical or drag load needs to be assumed in conjunction with the side load.

(B) **One-wheel landing** — An investigation of the fuselage structure is required for a one-wheel landing, in which only those loads obtained on one side of the fuselage in the level landing condition are applied. The resulting load factor is therefore one-half of the level landing load factor. This condition is identical with the level landing condition insofar as the landing gear structure is concerned. The unbalanced rolling moment will be resisted by angular inertia loads which are supplied largely by the wing structure. See Aeronautics Bulletin No. 26, Section 34, for additional information on this subject.

SEC. 38. BRAKED LANDING

(A) The applied load factor for this condition shall be 1.33.

(B) Airplanes equipped with brakes shall be investigated for the loads incurred when a landing is made with the wheels locked and the airplane is in an attitude such that the tail skid just clears the ground. The weight of the airplane minus the weight of the chassis shall be assumed to act on the wheels in a direction perpendicular to the ground line in this attitude. In addition, a component parallel to the ground line shall be assumed to act at the point of contact of the wheels and the ground, the magnitude of this component being equal to the weight of the airplane, minus chassis, times a coefficient of friction of 0.55. The tire in all cases shall be assumed to have deflected not more than one-quarter the nominal diameter of its cross section, and the deflection of the shock absorbers shall be the same as in level landing.

SEC. 39. SHOCK ABSORPTION REQUIREMENTS[1]

(A) The height of free drop, in inches, shall be 0.36 times the calculated stalling speed in miles per hour, except that it

Aerodynamically clean gears are the round tapered rod type (shown below with streamlined fairings added), or the flat spring steel gear shown bottom. Both gears were developed by famed speed pilot, Steve Wittman.

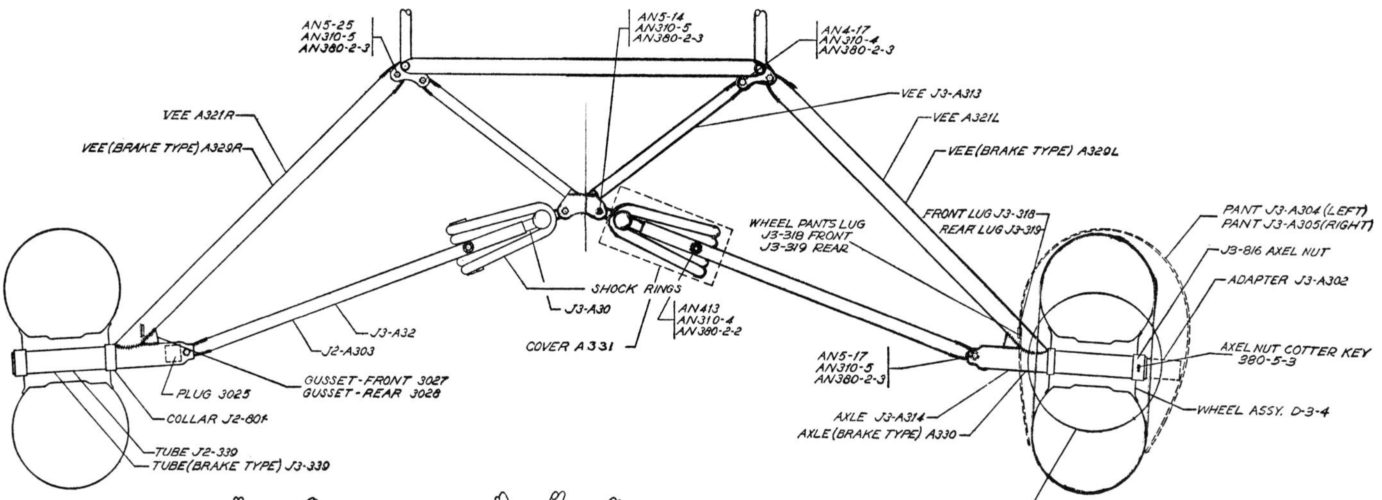

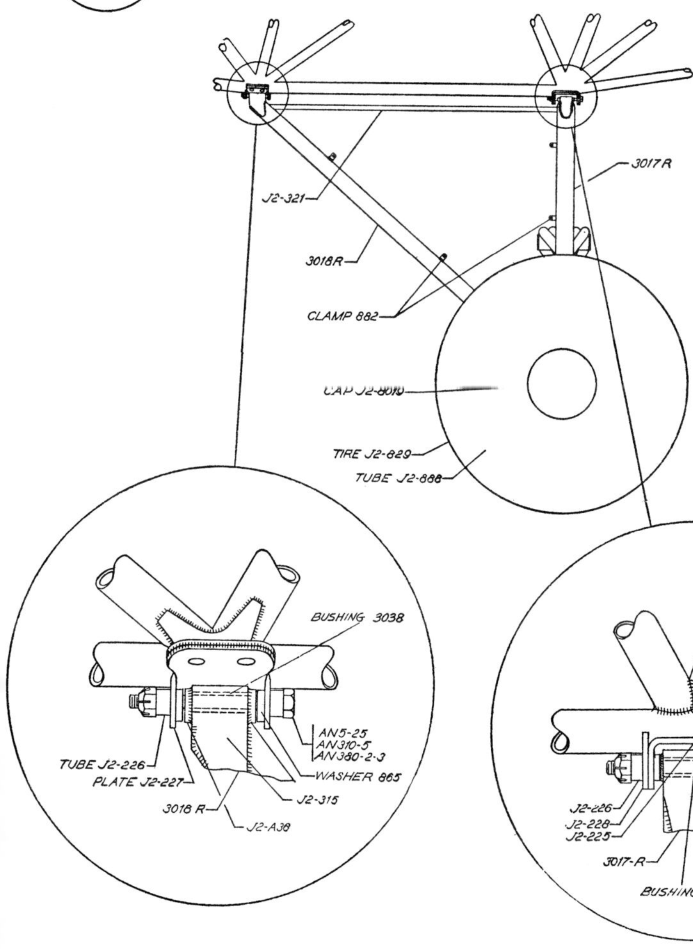

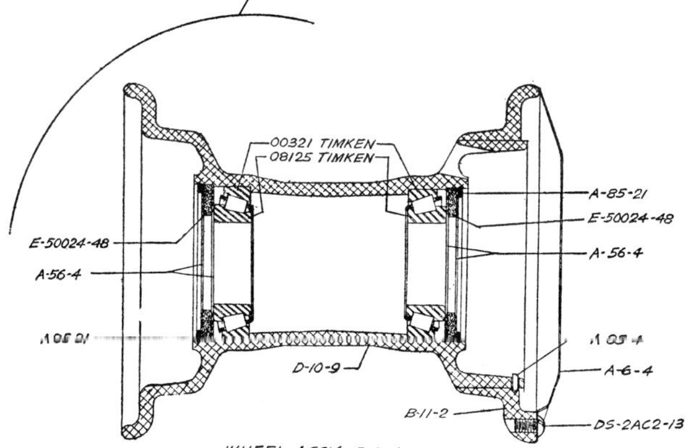

WHEEL ASSY. D-3-4

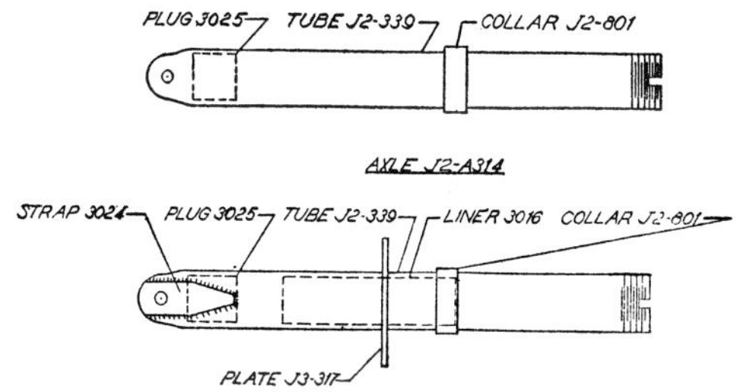

AXLE (BRAKE TYPE) A330

Piper J-3 landing gear is a simple, very practical design that is used on many homebuilt aircraft. Gear is extremely rugged, yet easy to maintain. Shock cords can be replaced quickly and easily. Most builders up date by replacing the original, and costly, 800 x 4 wheels with axles and wheels to fit 500 x 5 or 600 x 6 wheels with full disc brakes.

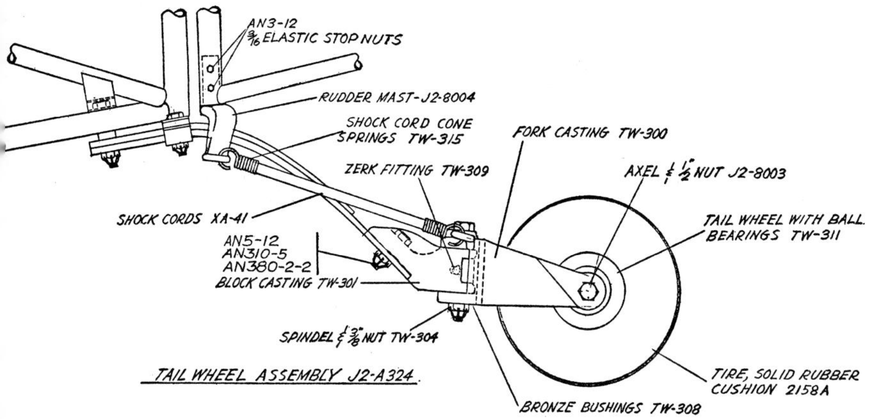

TAIL WHEEL ASSEMBLY J2-A324

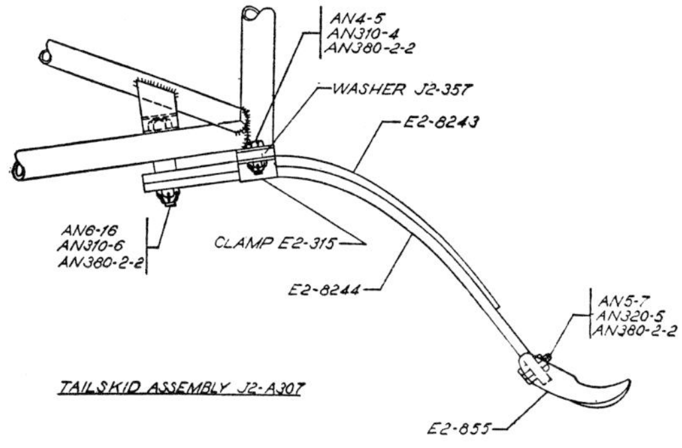

TAILSKID ASSEMBLY J2-A307

need not be greater than 18 inches for conventional airplanes. When auxiliary high-lift devices or "air brakes" are employed, the height of free drop shall not be less than 18 inches. (A special investigation of the energy to be absorbed may be required in the case of airplanes having exceptionally high sinking speeds, pending a general revision of shock absorption requirements to include basic airplane characteristics.) The height of free drop is measured from the bottom of the tires to the ground, with the landing gear extended to its extreme unloaded position. If the design of the shock-absorbing system is such the above method of specifying the required energy absorption capacity appears to give irrational results, an alternate method will be considered upon presentation of the pertinent data.

(B) The shock-absorbing systems, including that for the tail, shall be capable of absorbing the kinetic energy generated during the entire vertical travel of the airplane when the airplane is dropped from the required height of free drop, without subjecting any structural member to a load greater than its maximum design load. This applies to both the level landing and three-point landing conditions. The effects of support from the wings and the shock-absorbing capacity of the structure shall be neglected in determining the energy to be absorbed by the shock-absorbing system.

(C) The shock-absorbing units used in the shock-absorbing systems must conform to the requirements for approval outlined in Aeronautics Bulletin No. 7-F except that a drop test of the complete fuselage and landing gear assembly, under full-load conditions and with a suitable accelerometer or other satisfactory means of recording impact forces installed, will be acceptable as evidence of the shock-absorbing capacity of the strut.

SEC. 40. WHEELS

(A) Landing gear wheels shall be of a type and design approved by the Secretary in accordance with the requirements outlined in Aeronautics Bulletin No. 7-F and shall not be subjected to loading conditions in excess of those for which they are approved. The static working load of a landing gear wheel shall be taken as half the gross weight of the airplane upon which it is installed. The static working load for a tail wheel shall be taken as the maximum actual load which may be imposed on the tail wheel when the airplane is at rest on the ground.

(B) When the wheels installed on a specific airplane are of such a size as to require that a tire of special construction be used to support the airplane weight, **the wheels shall be plainly and conspicuously marked to that effect.** Such marking shall state the make and size of the proper tire and the number of plies required in its construction.

SEC. 41. SPECIAL CONSIDERATIONS

(A) The structural members of wire-braced landing gear shall be designed to provide a factor of safety of at least 1.5 against the rigging loads computed in accordance with section 62 (A), (B), and (C). For landing gear structures which are a part of a wing lift truss, section 62 (D) shall be complied with.

(B) The tail skid or tail wheel installation may be dropped from the height specified in section 39 in lieu of submitting a tail skid or wheel installation analysis, if the manufacturer so desires.

(C) When retractable landing wheels are used, means shall be provided for indicating to the pilot, at all times, the position of the wheels. Separate indicators for each side are required unless a single indicator is obviously satisfactory. A positive look shall be provided for the wheels in the extended position, unless a rugged irreversible operating mechanism is used.

(D) Care shall be taken to avoid stress concentrations in members subjected to dynamic loads, especially those which have been highly heat-treated. All fillets in members subjected to bending shall be made as large as is practicable.

(E) Landing gear members which also form part of a lift truss shall be designed in accordance with section 25 (F).

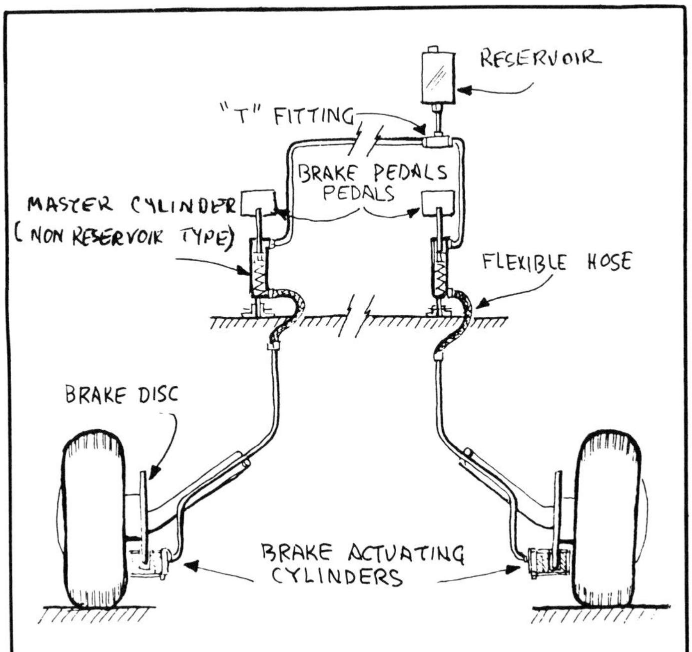

Simple Hydraulic System (Reservoir Type)

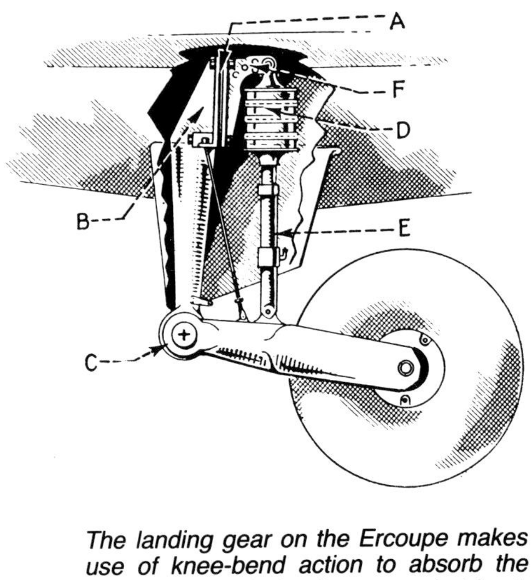

The landing gear on the Ercoupe makes use of knee-bend action to absorb the shock. The main landing gear strut is attached to the wing beam "A" by the fitting "B". This strut is hinged at "C" to allow for verticle movement of the wheel. The shock is taken up in the shock absorber "D" through the strut "E". This unit is mounted on the trailing edge of the wing beam of "F".

The retractable gear installed on the Celerity — a two place, low wing aircraft powered with an 0-320. Views above left show gear in retract position on partially completed aircraft, as well as above in extended position showing landing gear door details. Left shows retractable tail wheel assembly. Aircraft is constructed of foam and fiberglass with epoxy resin covering over a basic wooden airframe.

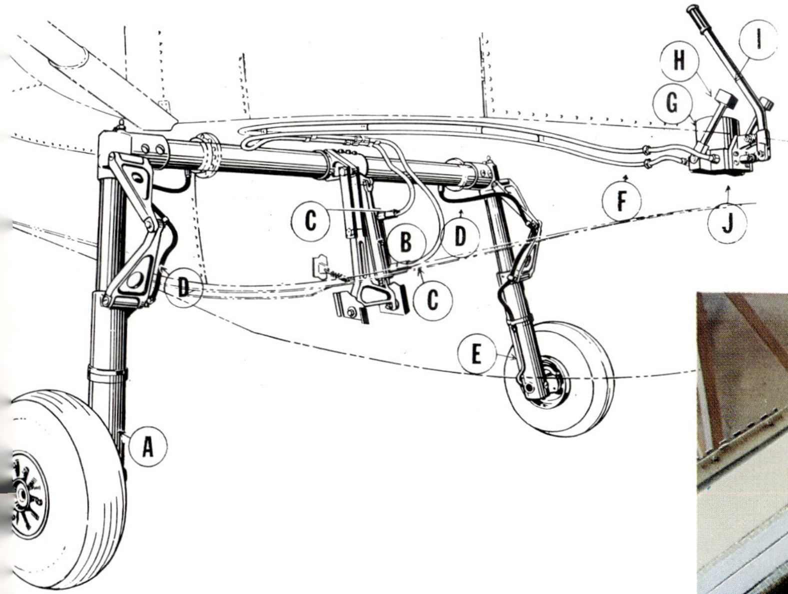

Drawing of the landing gear installation on the Republic Seabee. (A) Shock strut unit; (B) main landing gear cylinder; (C) pressure hose; (D) pressure hose; (E) wheel brake hydraulic line; (F) tubing; (G) selector valve unit; (H) wheel retraction handle; (I) hand pump; and (J) Electrol Powerpak.

Construction details of retractable landing gear mechanism of a Glasair III. Excellent workmanship displayed here was evident throughout the aircraft and was recognized by the judges at Sun 'n Fun EAA Fly-In 90 — winning Homebuilt Grand Champion honors for master builder Jerry Gruber — his second, having won a similar award in 1985 with a Glasair RG.

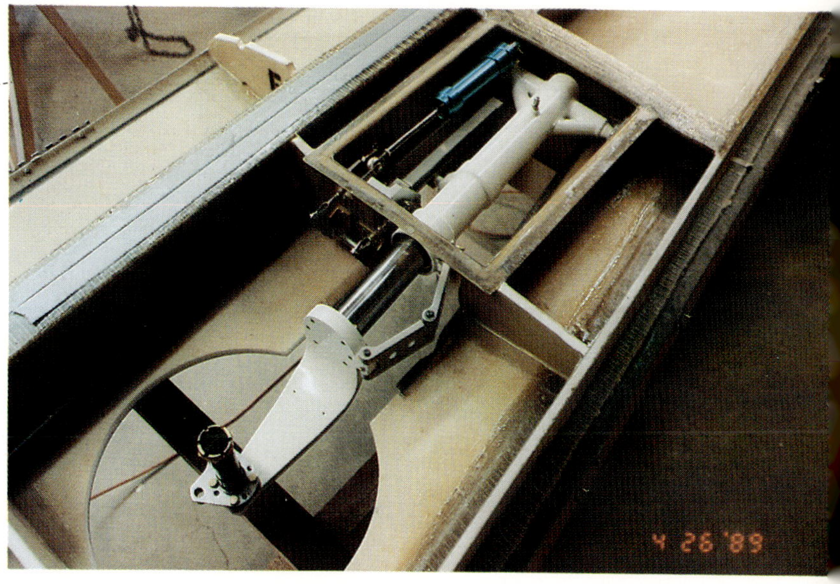

Landing Gear Notes

by Tony Bingelis

Sometimes in our eagerness to finish the airplane, we complete our final installations for the wheel pants, cowling, fairings, etc. with very little thought given to the need for future inspection and maintenance. Instead, we expeditiously make the installations as simple as possible and don't concern ourselves with such things as ease of access or long term service suitability.

A few of the notes I have gathered over the years regarding landing gear brakes, tires, and wheel pants installations, in particular, seems to bear my conclusion.

For example, I have for the past 10 years or so always installed my brake lines using Nylaflow tubing with brass automotive type compression fittings and connectors. This type of installation has given me excellent service for years in several different aircraft - including the retractable Falco.

Maybe the reason those installations worked so well is because I didn't habitually ride the brakes. I am not a lead footed brake user.

I have heard that a few builders have experienced brake problems because the Nylaflow tubing softened and pulled loose from its connector at the wheel brake assembly after a prolonged hard application of the brakes. This failure was undoubtedly due, in part, to the improper installation of the nylon brake line fitting (see Figure 1), and to the high heat generated from heavy brake applications.

Some aircraft, characteristically, require a greater use of brakes than others . . . and so do some pilots.

I think that most taildraggers, and tricycle gear aircraft with direct steering nose wheels are the least demanding because their brakes are used primarily during runups, taxiing, and for turning off the runway after landing. There are some exceptions, of course. A narrow landing gear, for example, or improperly aligned wheels and faulty landing techniques can increase the need for spirited if not excessive braking actions.

If you have a fast-landing aircraft, or a tricycle gear job with a free swiveling nose gear and not a steerable one, be prepared for an increased brake usage. Part of this is due to the need to maintain initial directional control with the brakes during the takeoff run until the rudder becomes effective with increased speed.

The RV-6A I am currently building falls into this category along with other aircraft like the VariEze, Grumman American Yankee, Sidewinder, and similar free swiveling nose wheel types.

That means I may have to consider the likelihood that my old favorite brake line hookup, using Nylaflow tubing all the way, should probably be modified to an installation that can tolerate a higher heat build-up at the wheel brake assembly.

BRAKE LINE OPTIONS

The brake line running up from the wheel brake assembly can be made to be more heat tolerant by installing a short length of flexible hydraulic hose at the wheel brake assembly. Or you could substitute aluminum tubing for the Nylaflow tubing.

If you use soft aluminum tubing for your brake line, it can be formed and bent to run up the landing gear leg all

Perhaps the best brake line installation results when a short length of medium pressure hydraulic hose is connected directly to the brake assembly. The upstream portion of the brake line can be either aluminum or Nylaflow tubing.

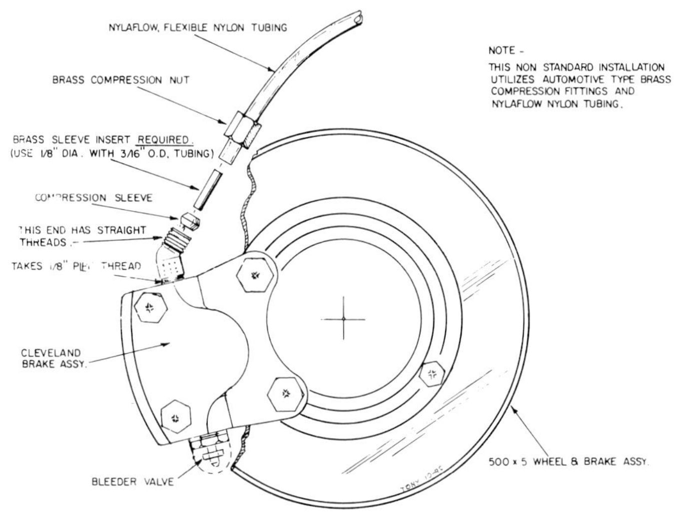

FIGURE 1.
NYLAFLOW BRAKE LINE INSTALLATION

Aluminum tubing is often used in much of the brake line system. However, it is more difficult to form, fit and route than are flexible aircraft hoses or Nylaflow tubing.

This entire brake system is made up of flexible, easy to route, Nylaflow tubing using automotive type compression fittings and connectors.

You can almost double the number of landings you will be able to get out of a pair of tires simply by rotating them periodically. Why do you suppose one tire has much more wear than the other?

the way to a bulkhead fitting on the firewall, or to the side of the fuselage if the need dictates. In other words, no other intermediate connection should be necessary.

However, unlike the flexible Nylaflow tubing, the bends in aluminum tubing are somewhat harder to form and route directly on the aircraft.

Perhaps the best installation at the wheel brake assembly is one made with a short length of a medium pressure aircraft hydraulic hose (Aeroquip 303 or similar). The Cleveland Wheels and Brakes folks recommend a minimum 6 inch length for this flexible portion of the brake line.

Such a hose is quite heat tolerant and its flexibility makes it easy to connect and route. Its one drawback is that it introduces an additional connection that can become another potential leak source. However, that is but a remote risk.

Connecting the end of the flexible aircraft hose to an aluminum line can be easily done using standard aircraft fittings. However, if you want to transition from the standard flexible aircraft hose to a Nylaflow line for the remainder of the upstream portion of the brake system installation, the job is a bit more complex. This is because you may have to use a mix of standard aircraft fittings and automotive type compression fittings (see Figure 2).

WHEEL PANTS CAN BE A PROBLEM

No doubt about it . . . wheel pants do give an airplane that classy finished look. And, for that matter, wheel pants, when properly designed and installed, can increase the performance of a clean fast airplane by as much as 5 mph . . . some builders claim even more.

Wheel pants do, however, introduce some operational difficulties. Highly effective wheel pants cover so much of the wheel that it is difficult to inspect the tires for wear.

For this reason, the installation of your wheel pants should be carefully worked out so that their removal is relatively easy. Otherwise, you will be most reluctant to take them off to inspect your tires for unusual or excessive wear and damage. Likewise, your wheel and brake installation may be deprived of the attention it deserves.

You will find the need for easy access to the wheels and brakes will become far more important to you after you have the airplane flying than it was during construction.

IS YOUR BRAKE ASSEMBLY VULNERABLE?

As you look at your brake installation now, you may be struck with the sudden realization that the brake housing can be extremely vulnerable to damage or total destruction in the event of a flat tire or blow out. This will surely happen if you position the brake assembly below the axle rather than in front or behind it where it would be somewhat protected by the wheel rim. In some retractable installations it may be difficult, or impossible, to rotate the brake assembly to a safer position. If that be the case, be aware that a simple flat tire during landing or takeoff can result in far greater damage than a better positioned brake assembly would suffer.

TIRE VALVE ACCESS

Wheel pants, in addition to interfering with the inspection of the tires, make servicing them very difficult.

Getting air into the tires when they are low may require the removal of the wheel pants unless you have made some provision for access to the tire's valve stem. This can be very annoying when you want to go fly but instead have to waste time removing the pants to simply squirt a little air in the tires.

A few builders do anticipate the need to air the tires periodically and wisely provide small access doors in the wheel pants for this purpose. These access doors are fine and serve the purpose nicely. However, they do require a lot of work to fabricate and install.

A far simpler way to access the tires is illustrated in Figure 3.

All you have to do is obtain a large diameter stainless steel button plug (cap?) from a hardware store, cut a hole in the wheel pants, and press the plug in.

I use a 1-1/4" chrome plated button and cut the opening in the wheel pants for it with a hole saw. The entire installation cost less than $2 and took less than 10 minutes to install.

If you don't like the appearance of large shiny chrome tire access buttons in your wheel pants, you could, of course, paint them to match the wheel pants color. Then they will not be at all conspicuous.

By locating the access hole in the wheel pants at a level below that of the axle, you will be able to reach under the wheel pants opening and pop the button out with your fingers for servicing.

Be sure to locate the hole far enough away from the axle nut to clear the button's tabs when installed. This dimension will depend on the size of the wheels you have installed.

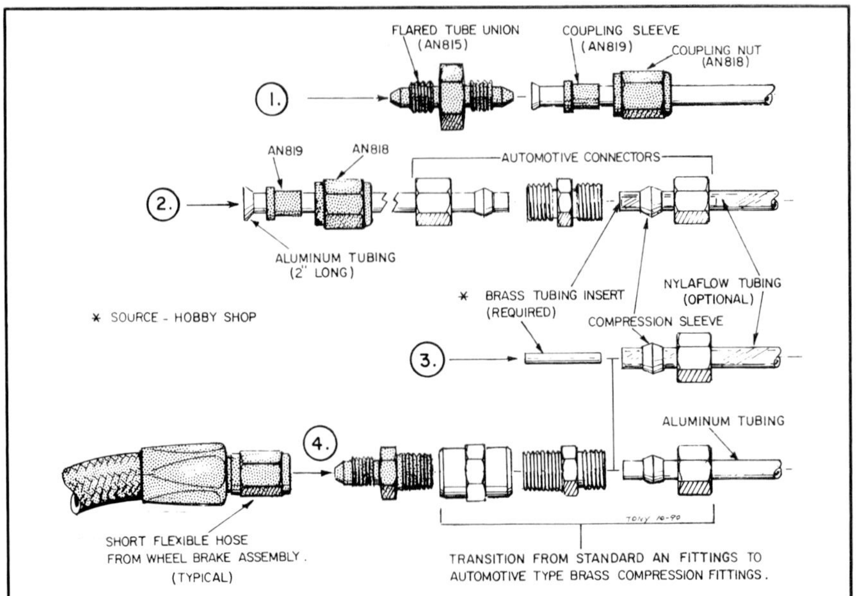

FIGURE 2. BRAKE LINE OPTIONS

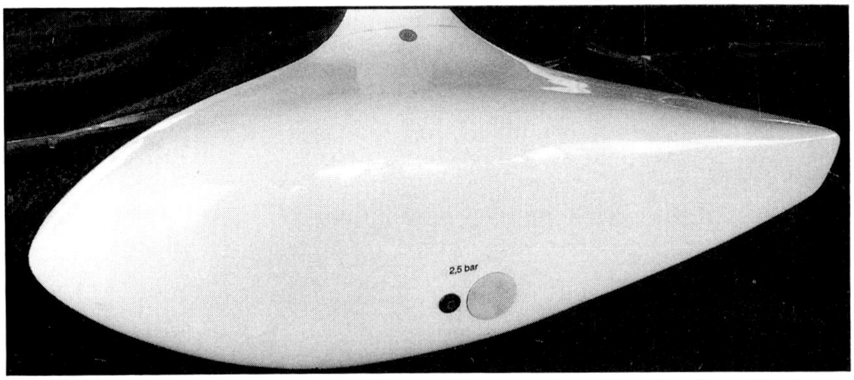

A large hole in the wheel pants provides quick access for airing the tires. The large diameter chrome plated button simply snaps into place to hide the hole.

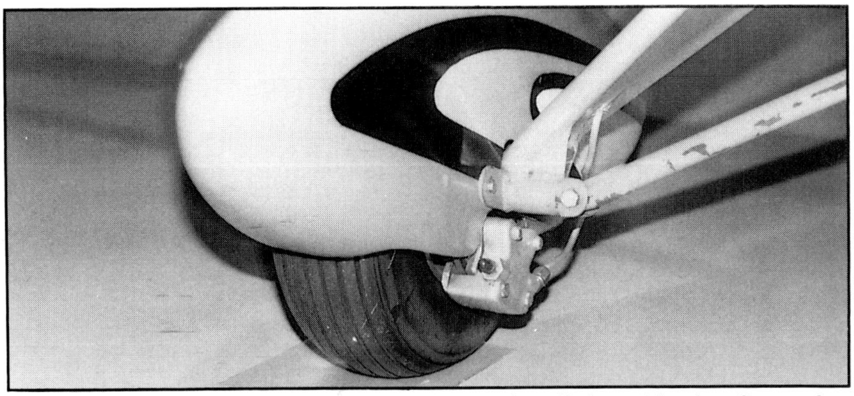
Brake assemblies protruding below the axles, as installed on this aircraft, may be damaged or totally destroyed in the event of a flat tire, or blowout, during takeoff or landing.

Right — Highly streamlined wheel pants cover so much of the wheel that it is difficult to inspect the tires for wear. Such wheel pants are also highly vulnerable when operating from unpaved airstrips.

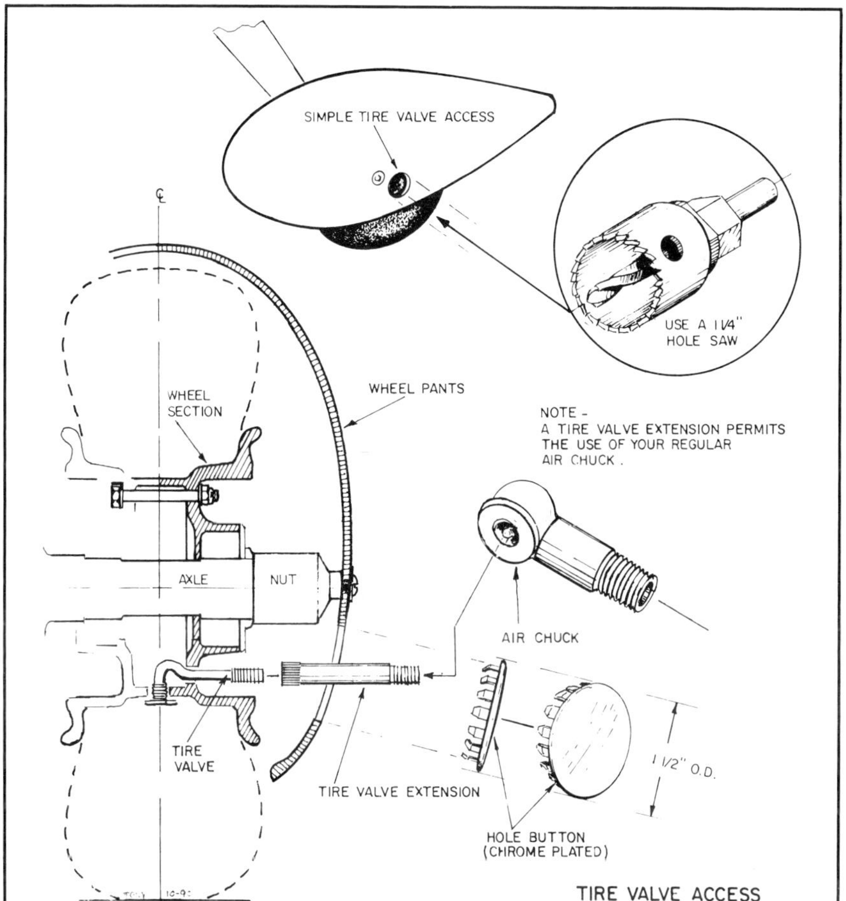

FIGURE 3. TIRE VALVE ACCESS FOR WHEEL PANTS

Remember, when you rotate the wheel so the valve stem is visible, it should be positioned in the opening so you can reach it with the air chuck.

To get an air chuck into that small opening will require one of two mechanical aids.

1. A long straight air chuck available at most any auto parts store for about $7. Or . . .

2. A tire valve extension. Although the extensions come as a set of 4 (about $3.75), all you will need is one unless your airplane is a four wheeler.

All you really need is to temporarily screw one of the extensions into the tire valve. It may permit you to use your regular air chuck. After airing the tire, simply unscrew the extension and stash it away in the aircraft in your emergency tool kit (you should have one, you know).

Since four of these little jewels come in a package, you might consider putting each one away in a different place. Then, when you need one, you might be lucky enough to remember where to look.

FOR INCREASED TIRE LIFE

Rotating the tires on your car allows you to get more mileage from them before being forced to buy a new set. Well, that practice is equally effective with aircraft tires.

You can almost double the number of landings you will be able to get out of a pair of tires simply by rotating them periodically . . . say, during your annual inspection.

Unfortunately, rotating the tires is a time consuming job because you will have to jack up the airplane and remove the wheel pants and the wheels. Then, I am sure you realize, you will have to remove both tires from the wheels, turn them around, and switch them to the opposite wheels. That is a lot of work but well worth the effort.

If you are like many pilots, you probably can take the blame for causing one of your tires to wear more than the other. It's that tendency some of us have of touch down on that wheel first - maybe it's because we habitually peek out the left (right?) side during landing?

Here's something else I consider. Because I am the one who has to buy the tires, I use full flaps whenever I can and try to 3-point it (preferably once) as slowly as possible. Making "wheelies" is a costly practice.

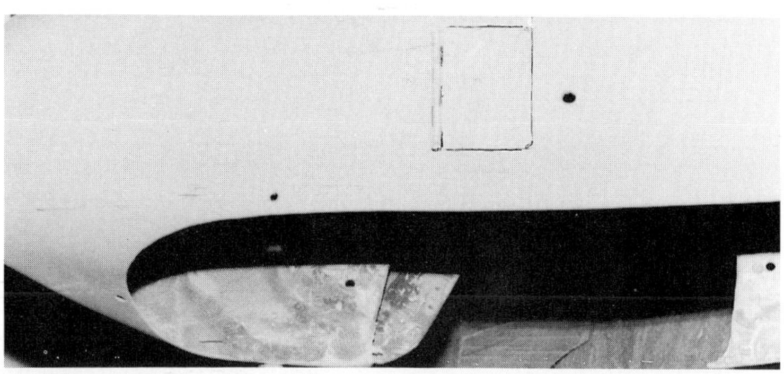

Some builders wisely provide small access doors in the wheel pants to permit airing the tires without having to remove the pants.

If you wish to contact the author of this column, **Sportplane Builder,** for additional information, please write to Tony Bingelis, 8509 Greenflint Lane, Austin, TX 78759.

Powerplant Installations

16-1 CERTIFICATED-TYPE AIRCRAFT ENGINES

The power plant installation is considered to include all components of the aircraft that are necessary for propulsion. It should also be considered to include all components which affect its safety of operation between normal inspection or overhaul periods.

Any certificated type aircraft engine or propeller combination, whether or not maintained in its certificated condition, may be used, provided that no adverse characteristics of the engine, propeller, or engine-propeller combination is evident or known to the builder or to the FAA.

16-2 NON-AIRCRAFT ENGINES

Any non-aircraft engine may be installed and used if suitable.

16-3 POWER REQUIREMENT

With take-off power at sea level, a steady rate of climb not less than 300 ft/minute should be possible.

16-4 INDUCTION SYSTEM

The engine air induction system must supply an adequate quantity of air to the engine under all conditions of operation.

16-5 INDUCTION SYSTEM DUCTS

Induction system ducts should be so installed as to prevent the accumulation of fuel or moisture in all normal ground and flight attitudes.

Drains should not discharge in a location which will constitute a fire hazard.

Ducts which are connected to components of the aircraft between which relative motion may exist should incorporate provisions for flexibility.

16-6 INDUCTION SYSTEM SCREENS

Where induction system screens or filters are employed, they should be located upstream from the carburetor. It should not be possible for fuel to impinge upon the screen. An alternate air intake or supply should be available in the event the screen becomes obstructed. It may also be used to furnish hot air to the carburetor when needed to melt ice or prevent its formation.

16-7 EXHAUST SYSTEM

The exhaust system should be constructed and arranged in such a manner as to assure a safe disposal of exhaust gases without the existence of a fire hazard or contamination of the air in the cockpit area with carbon monoxide.

Unless other suitable precautions are taken, all exhaust system components should be separated from adjacent flammable portions of the aircraft by means of fireproof shields.

Exhaust gases should not be discharged within dangerous proximity of any fuel or oil system drains.

All exhaust system components should be ventilated to prevent the existence of points of excessively high temperature.

16-8 EXHAUST MANIFOLDS

While exhaust manifolds for certificated-type aircraft engines may be fabricated of automotive-type tubes, preference should be for the use of fireproof, corrosion-resistant materials, with incorporated provisions designed to prevent failure due to their expansion when heated to operating conditions.

Exhaust manifolds are to be supported in a manner adequate to withstand all vibration and inertia loads to which they might be subjected in operation.

Portions of the manifold which are connected to components between which relative motion might exist should incorporate provisions for flexibility.

16-9 HEAT EXCHANGES (HEAT MUFFS)

Each exhaust heat exchanger or heat muff must be constructed and installed to withstand the vibration, inertia, and other loads that it may be subjected to in normal operation.

Each exchanger must be suitable for continued operation at high temperatures and should be resistant to corrosion from exhaust gases.

There should be means for inspection of critical parts of each heat exchanger.

Each heat exchanger should be ventilated where it is

Six cylinder Franklin installation in a Great Lakes replica. Very compact but neat installation. Part of inverted oil system is the three hole unit shown extreme right on firewall.

Continental Model 0-200-A Aircraft Engine

Compression Ratio: 7.0:1
Displacement: 201 cu. in.
Fuel: 80/87 grade min.
Power corrected to 29.92 In. Hg.
60°F. Inlet Air Temp.

Oil Required	Below 40°F: SAE #20
	Above 40°F: SAE #40
Oil Temperature	75°.—170°F.
Oil Pressure, Idling	10 p.s.i. Min.
Oil Pressure, Cruising	30-60 p.s.i.
Oil Sump Capacity	6 quarts
Allowable Temperatures:	
Cylinder Head	525°F. Max.
Cylinder Barrel	290°F. Max.
Oil Temperature	225°F. Max.

Type Certificate Number	252
Number of Cylinders	4
Bore (Inches)	4.06
Stroke (Inches)	3.88
Displacement (cubic inches)	201
Type of Propeller Drive, Flanged	Direct
Compression Ratio	7.0:1
Rated Power (Sea Level)	100
Take-Off Power (Sea Level)	100
Recommended Cruise Rating (Sea Level)	75
RPM at Rated Power	2750
RPM at Take-Off	2750
Cruising RPM	2500
Recommended Fuel Grade Min.	80/87
Dry Weight (With Accessories Below Included)	217.87
Crankshaft Rotation	Clockwise
Ignition	Dual
Dimensions With Standard Equipment Installed — Length	28.53
Width	31.56
Height	28.72

STANDARD EQUIPMENT

ITEM	MAKE	WEIGHT
Spark Plugs	A.C.	1.75
Magnetos	Slick	12.12
Ignition Harness (Shielded)	Continental	3.81
Starter	Delco-Remy	15.50
Generator	Delco-Remy	10.12
Oil Cooler	Harrison	4.25
Tachometer Drive	Continental	0.12

subject to contact with exhaust gases.

Any heat exchanger used for heating cockpit or cabin ventilating air must be constructed so that exhaust gases may not enter the ventilating air.

16-10 COOLING

The powerplant cooling provision should be capable of maintaining the temperature of all powerplant components, engine parts and engine fluids (oil and coolant), at or below the maximum established safe values under critical conditions of ground, takeoff, and climb operation.

16-11 IGNITION SYSTEMS

Each battery ignition system must be supplemented by a generator that is automatically available as an alternate source of electrical energy to allow continued engine operation if any battery becomes depleted.

The capacity of batteries and generators should be large enough to meet the simultaneous demands of the engine ignition system and the greatest demands of any electrical system components that draw from the same source.

16-12 POWERPLANT ACCESSORIES

Engine-driven accessories should be of a type satisfactory for installation on the engine involved and ordinarily would utilize the provisions made on the engine for the mounting of such units.

Items of electrical equipment subject to arcing or sparking should be installed so as to minimize the possibility of their contact with any flammable fluids or vapors which might be present in a free state.

16-13 POWERPLANT CONTROLS

Location of powerplant controls should be such as to provide convenience in operation including provisions to prevent inadvertent operation.

The controls should be located and arranged so that when the pilot is seated it will be readily possible for him to obtain full and unrestricted movement of each control without interference from either the pilot's clothing or the cockpit's structure.

Controls should maintain any set position without constant attention by the pilot and should not tend to creep due to control loads or vibration.

Flexible controls should be of a suitable type for the function to be performed.

Controls should have adequate strength and rigidity to withstand loads without failure or excessive deflection.

16-14 THROTTLE CONTROLS

A throttle control should be provided to give independent control for each engine. Throttle controls should afford a positive and immediately responsive means of controlling the engine. VW and other engines, spring load throttle to open, not close, in event of throttle failure.

16-15 MIXTURE CONTROLS

If a mixture control is provided, a separate control should be installed for each engine.

The controls should be grouped and arranged in such a manner as to permit both separate and simultaneous control of each engine.

16-16 IGNITION SWITCHES

Ignition switches should provide control for each ignition circuit on each engine.

16-17 CARBURETOR HEAT

The hot air system to the carburetor, if the engine type requires one, should be such that a minimum temperature rise of 90°F is attainable with sea level engines and conventional venturi carburetors when the outside air temperature is 30°F and the air free from visible moisture.

Pressure injection-type carburetors on sea level engines when equipped with carburetor heat controls, although less subject to icing, should provide a heat rise of approximately 60°F.

16-18 DIRECTION OF ENGINE CONTROLS MOVEMENT

The direction of movement of controls should be as follows:

THROTTLE: Forward to increase forward thrust.
PROPELLER: Forward to increase rpm.
MIXTURE: Forward or upward for rich.
CARBURETOR HEAT: Forward for cold air.

16-19 INSTALLATION AND TESTING

All components of the powerplant installation should be constructed, arranged and installed in a manner which will assure their continued safe operation between normal inspections or overhaul periods.

Accessibility should be provided to permit and facilitate inspection and maintenance as necessary to assure continued airworthiness.

The complete powerplant installation including the propeller, as installed in the aircraft, should satisfactorily undergo at least one hour of trouble-free ground operation from idling to full throttle power prior to the first flight. The operator may use any time interval he desires at the various speeds he selects.

The engine manufacturer's advice regarding run-in time as described in the engine operation manual, should be followed.

16-20 TESTING OF NON/AIRCRAFT ENGINES

All engines not originally and specifically designed for aircraft use, including essential accessories, should be subjected to ground tests and inspections to establish operational reliability.

The non-aircraft engine should be subjected to at least four hours of trouble-free ground operation from idling to full throttle and whatever calibration tests necessary to establish its power characteristics and reliability. The results of those ground tests and the subsequent longer mandatory flight test period required for non-aircraft engines should constitute the basis for establishing the characteristics of the engine over its entire operating range of crankshaft rotational speeds, manifold pressures, fuel/air mixture settings, altitudes, and unusual attitudes. The engine manufacturer's operating limitations should not be exceeded.

16-21 COOLING TESTS

A cooling test, to determine that the maximum temperatures permitted for the engine will not be exceeded, should be conducted by starting and running the engine on the ground until normal engine warmup temperature has been stabilized. The aircraft should then proceed with normal takeoff and climb at best-rate-of-climb speed. The power used should be takeoff power and speed and maximum continuous power and speed. The climb should continue for approximately five minutes after the temperature peak or dropoff, to assure that stabilization had been reached. This temperature should not exceed the maximum limit established by the engine manufacturer.

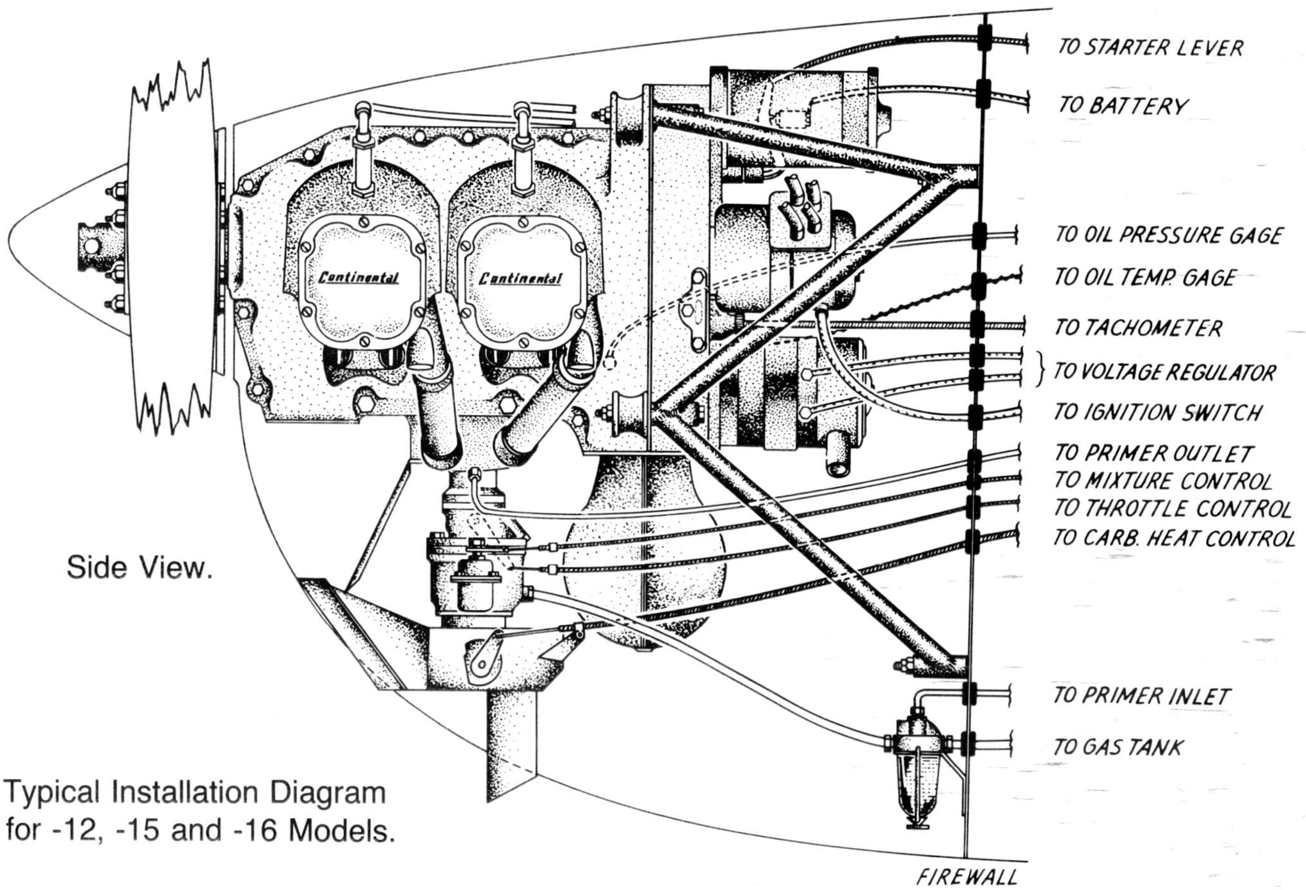

Typical Installation Diagram for -12, -15 and -16 Models.

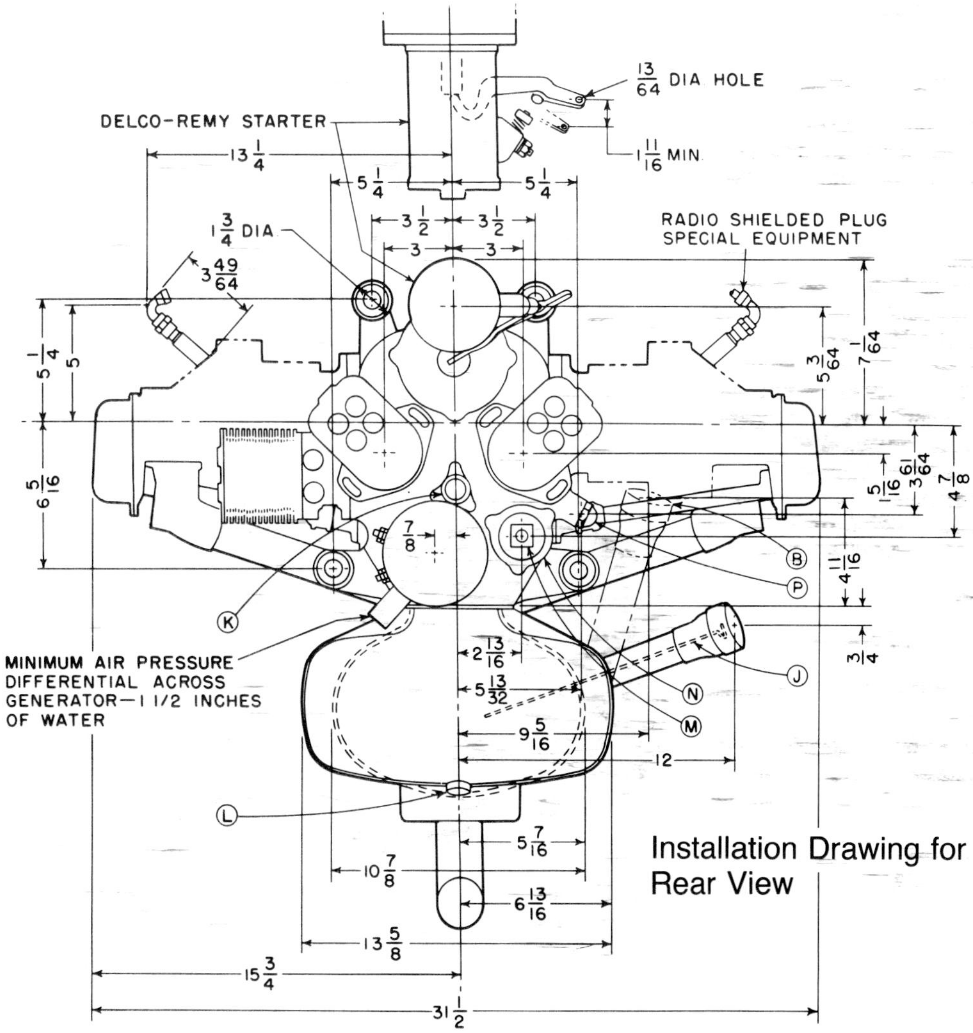

Installation Drawing for -12, -14 and -16 Models.
Rear View

Continental Aircraft Engine

A. Name plate
B. Oil filler (optional location)
C. S.A.E. No. 1 flanged crankshaft (optional)
D. Fuel pump - special equipment
E. Breather connection ⅝ I.D. hose
F. Exhaust mounting
G. Throttle contol lever
H. Oil pressure gauge connection
J. Oil quantity gauge
K. Tachometer drive S.A.E. standard ½ engine speed C.W. rotation
L. Oil drain ⅝-18 N.F. tap and plug
M. Oil temp. capillary ⅝-18 N.F. tap
N. Oil screen
P. Oil pressure relief valve
R. Air intake housing
S. Delco-Remy generator
T. Delco-Remy starter
U. Oil cooler optional special equipment

16-22 COWLING

Each cowling should be constructed and supported so that it can resist any vibration, inertia, and air loads to which it may be subjected in operation.

There must be means for rapid and complete drainage of each area of the cowling in the normal ground and flight attitudes.

No drain may discharge where it will cause a fire hazard.

The cowling should be at least fire resistant.

Each part behind an opening in the engine compartment cowling should be at least fire resistant for a distance of at least 24 inches aft of the opening.

Any part of the cowling subjected to high temperatures due to its nearness to exhaust system ports or exhaust gas impingement, should be fireproof.

16-23 PROPELLERS

Propellers installed in amateur-built aircraft should be a type which will properly restrain the engine to a speed not exceeding its maximum permissible take-off speed during take-off and initial climb at best-rate-of-climb speed with the engine operating at full throttle, or the throttle setting corresponding to maximum permissible take-off manifold pressure.

Fixed-pitch wood propellers of a particular pitch, diameter, and manufacture should withstand, without failure, two (2) hours of cumulative operations at take-off rpm.

Fixed-pitch metal propellers of a diameter and pitch range previously recommended by the propeller manufacturer for any particular engine-propeller combination are considered as acceptable.

Adjustable-pitch propellers determined as satisfactory for any given engine-propeller combination are considered as acceptable.

16-24 PROPELLER VIBRATION AND CUT DOWN PROPS

Metal propellers cut down and repitched beyond the propeller manufacturer's recommended limits may be subjected to such additional inspections and tests as requested by the FAA.

Caution

Unless proper vibration investigations have been made on such cut-down metal propellers installed with specific engine types and models, the likelihood is that fatigue failures of the blades may occur without prior indications.

For multi-engine and other high performance airplane configurations the propeller manufacturer should be requested to assist in determining that the propeller-engine combination has satisfactory vibration characteristics.

Evidence of satisfactory vibration investigations for fixed pitch metal and constant speed propellers with metal blades for a particular engine model is shown on the FAA Propeller Specifications for single engine tractor-type airplanes.

16-25 PROPELLER CLEARANCE

Ground clearance should be not less than seven inches in taxi and take-off attitudes.

Water clearance should be not less than 18 inches between each propeller and the water.

A one inch radial clearance should be provided between the blade tips and the aircraft structure, or whatever additional radial clearance is necessary to preclude harmful vibration of the propeller or aircraft.

Positive clearance should be provided between any other rotating parts of the propeller and stationary portions of the aircraft.

16-26 FIREWALLS

A suitable means, consistent with the size and complexity of the aircraft, should be provided to reduce the hazard of fire in the engine compartment.

Therefore, all engines and other combustion equipment which are intended for operation in flight must be isolated from the remainder of the aircraft by means of firewalls, shrouds, or other equivalent means.

16-27 FIREWALL CONSTRUCTION

a. Firewalls and shrouds should be constructed so that no hazardous quantity of liquids, gases, or flame can pass from the engine compartment to other portions of the aircraft.

b. All openings in the firewall or shroud should be sealed with fireproof grommets, bushings, or firewall fittings.

16-28 FIREPROOF MATERIAL TEST

The test for demonstrating that a material or unit is fireproof as outlined by the FAA subjects the material or unit to a 2000° (plus or minus) 50°F. flame. Sheet materials are tested by subjecting a sample approximately 10 inches square to a flame from a suitable burner. The flame must be large enough to maintain the required test temperature over an area approximately five inches square. The fireproof firewall materials and fittings must resist flame penetration for 15 minutes.

16-29 ACCEPTABLE FIREWALL MATERIALS

FAA considers the following materials satisfactory as firewalls and shrouds without testing:

a. Stainless steel sheet, .015 in. thick.

b. Low carbon sheet steel .018 thick coated with aluminum or protective coating.

c. 26 gauge galvanized sheet.

d. Terne plate, .018 in. thick.

e. Monel metal, .018 in. thick.

f. Steel or copper base alloy firewall fittings.

g. Iconel (Nickel-Chromium-Iron Alloy) sheet, .015 in. thick.

Wooden firewalls, lined with thin asbestos, as used in the early days, are unsatisfactory. They would, however, be acceptable when overlayed with one of the materials listed above.

Where the engine, during operation, is situated so that the probable path of an engine fire would not contact the aircraft structure, firewalls and shrouds may be constructed of fire-resistant materials . . . that is materials that do not support combustion.

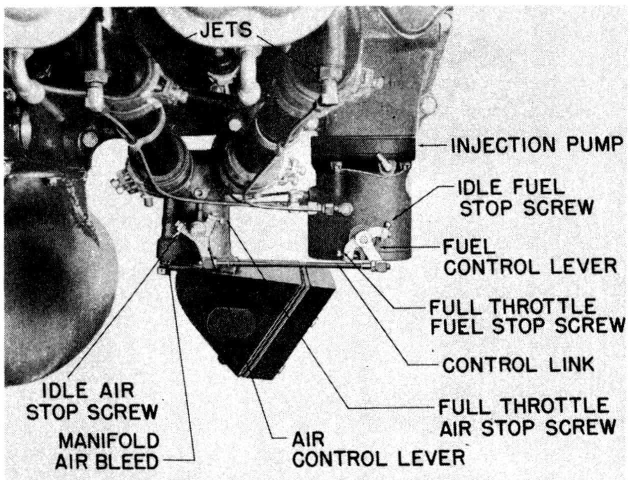

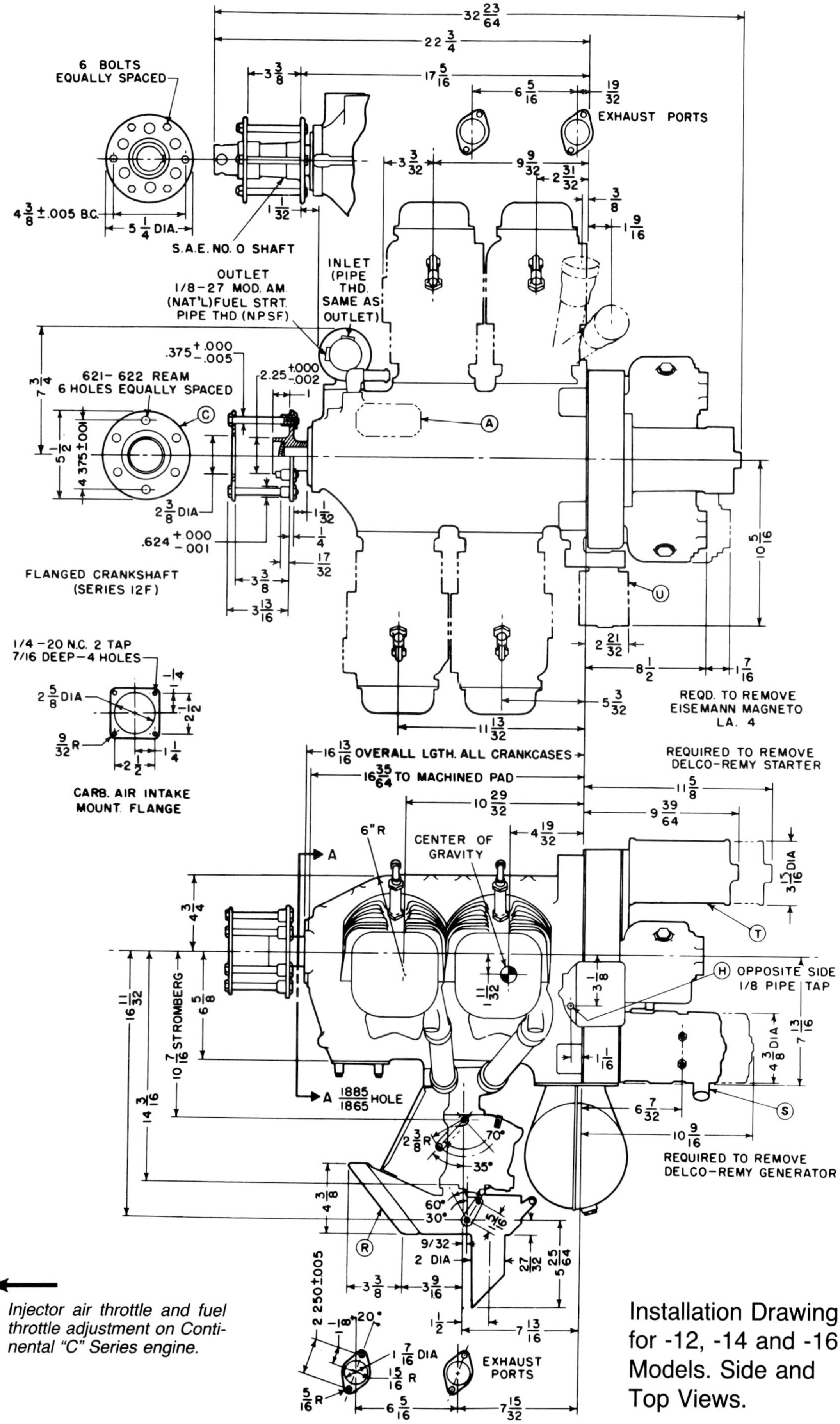

Injector air throttle and fuel throttle adjustment on Continental "C" Series engine.

Installation Drawing for -12, -14 and -16 Models. Side and Top Views.

Lycoming 0-320-B1A

Model	0-320-B1A, -B2A*
C.A.A. Type Certificate	274
Rated Horsepower	160
Rated Speed, RPM	2700
Cruise Speed	2100
Fuel Consumption, Cruise, GPH @	
75% Rated Power	10.0
65% Rated Power	8.8
60% Rated Power	8.2
Propeller Drive Ratio	1:1
Propeller Shaft Rotation	Clockwise
Bore, Inches	5.125
Stroke, Inches	3.875
Displacement, Cubic Inches	319.8
Compression Ratio	8.5:1
Dry Weight (With Starter & Generator). Lb.	278
Dimensions: Height, Inches	23.12
Width, Inches	32.24
Length, Inches	29.56
Oil, SAE Number, above 40°F.	50
below 40°F.	30
below 10°F.	20
Oil Sump Capacity, Quarts	8
Fuel, Aviation Grade, Octane	91/96
Carburetor, Marvel-Schebler	MA-4SPA
Magnetos (2), Scintilla	S4LN-20 (Right) S4LN-21(Left)
Tachometer Drive, Ratio to Crankshaft	0.5:1
Rotation	Clockwise
Starter Drive, Ratio to Crankshaft	13.55:1
Rotation	Counter-Clockwise
Starter, Delco-Remy, Geared 2.27:1 with Switch	Standard
Propeller Governor Drive, AN20010 Type XX Ratio to Crankshaft & Rotation	.866:1 - Clockwise

*Model 0-320-B2A is similar to Model 0-320-B1A but does not have provisions for controllable propeller.

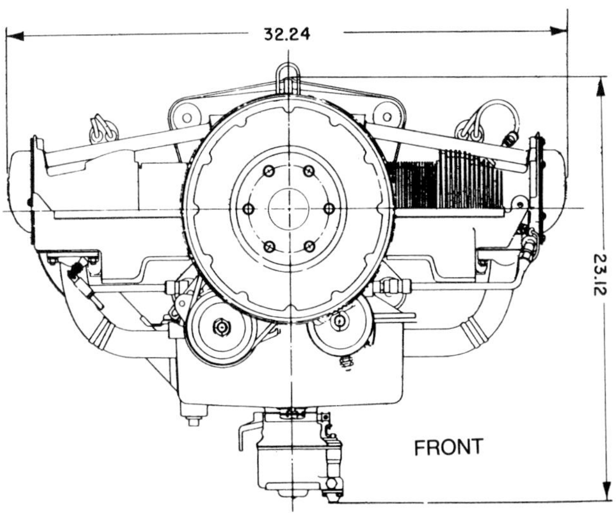

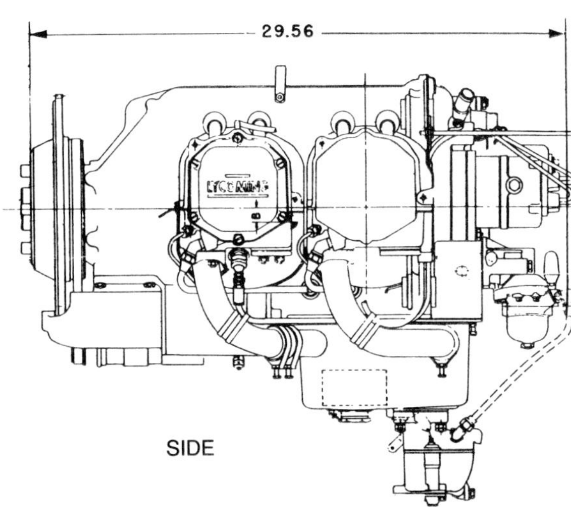

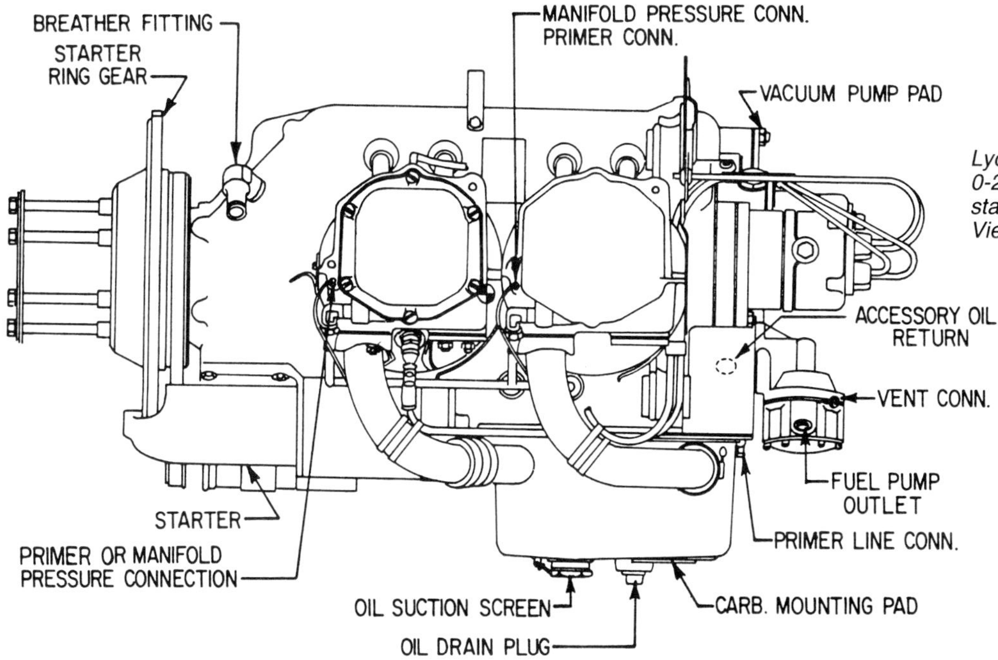

Lycoming 0-235-C, 0-290-D and 0-290-D2 Series engines — Installation Drawing — Left Side View.

Generator Drive,
  Ratio to Crankshaft 1.91:1
  Rotation . . . . Clockwise
Generator, Delco-Remy
  (12 V., 20 Amp.) . . . . . . Standard
Generator, Delco-Remy
  (12 V., 35 Amp.) . . . . Optional
Generator, Delco-Remy
  (24 V., 15 Amp.) . . . . Optional
Generator, Delco-Remy
  (24 V., 25 Amp.) . . . . Optional
Generator Voltage Regulator,
  Delco-Remy . . . . Standard
A-N Vacuum Pump Drive, Ratio to
  Crankshaft &
  Rotation . . . . 1.30:1 - C'clockwise
Vacuum Pump . . . . Optional
Optional Dual Drives Mounting on
  Vacuum Pump Drive Pad
  Vacuum Pump - Hydraulic Pump
    Drive, Ratio &
    Rotation . . 1.300:1 - C'clockwise
  Vacuum Pump - Propeller Governor
    Drive, Ratio &
    Rotation . 1.300:1 - C'Clockwise
Optional Fuel Pump Drives:
  Fuel Pump Drive, AN20003 Type
  X111-A, Ratio &
  Rotation . . . . .866:1 - C'clockwise
    or
  Fuel Pump Drive, Diaphragm
    Type . . . . Optional
Fuel Pump, AC . . . . Optional

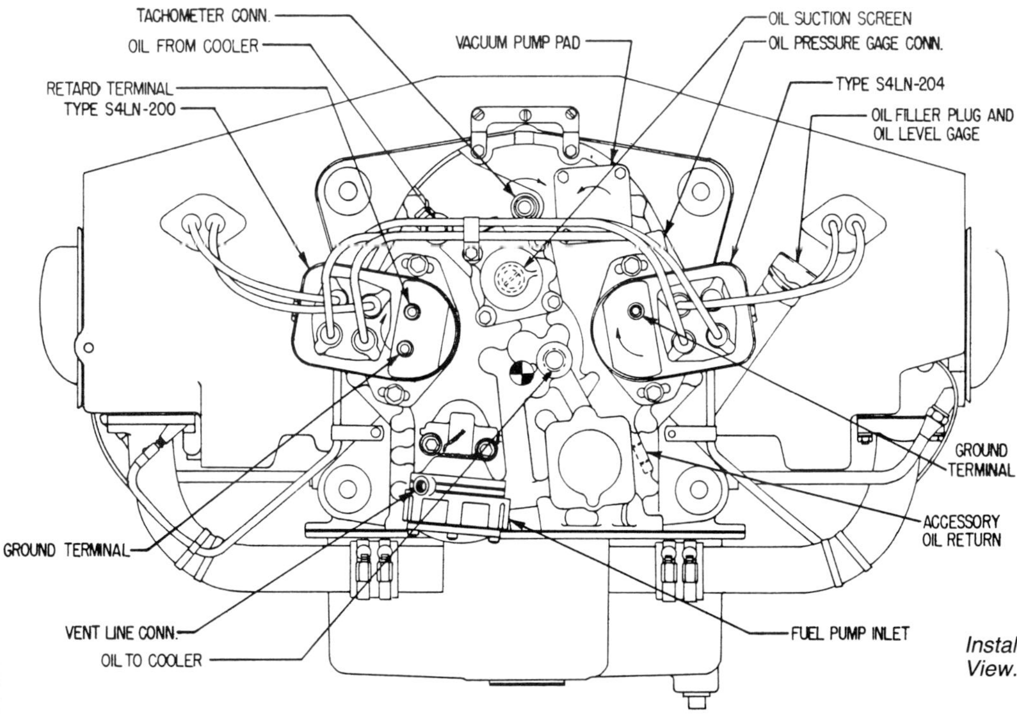

Installation Drawing — Rear View.

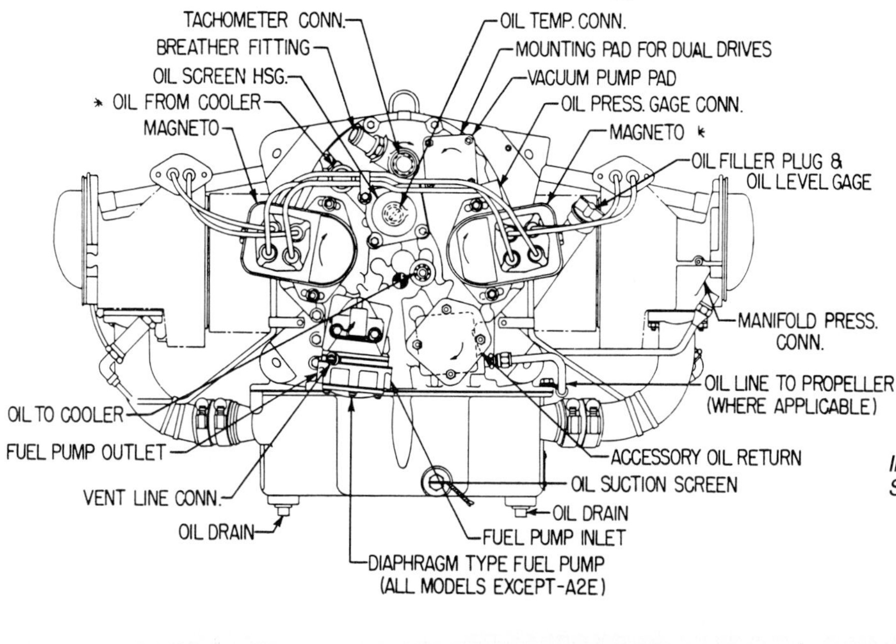

Installation Drawing — 0-360-A, -B Series (Excepting -A1C, -A1G).

VW engine installation on Pober Pixie.

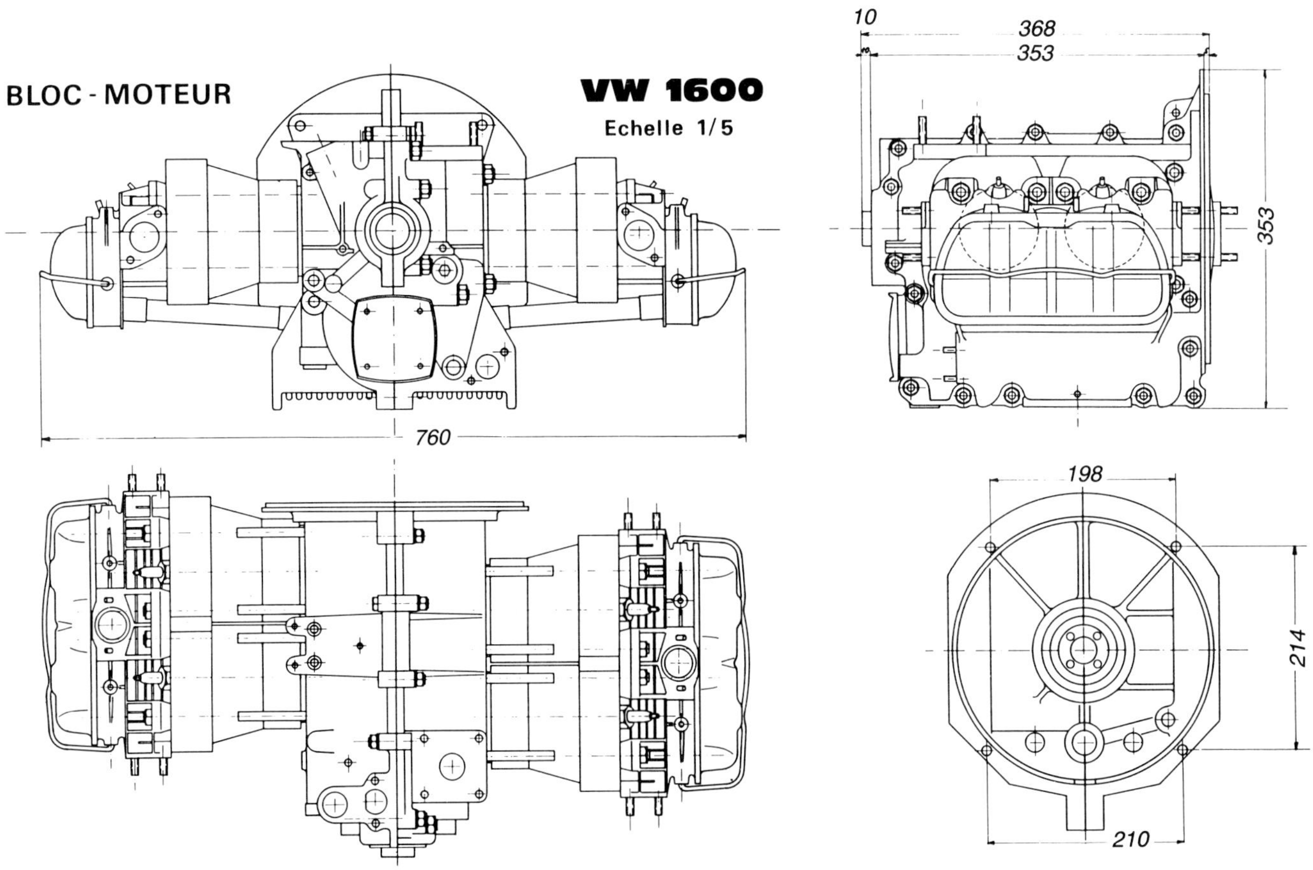

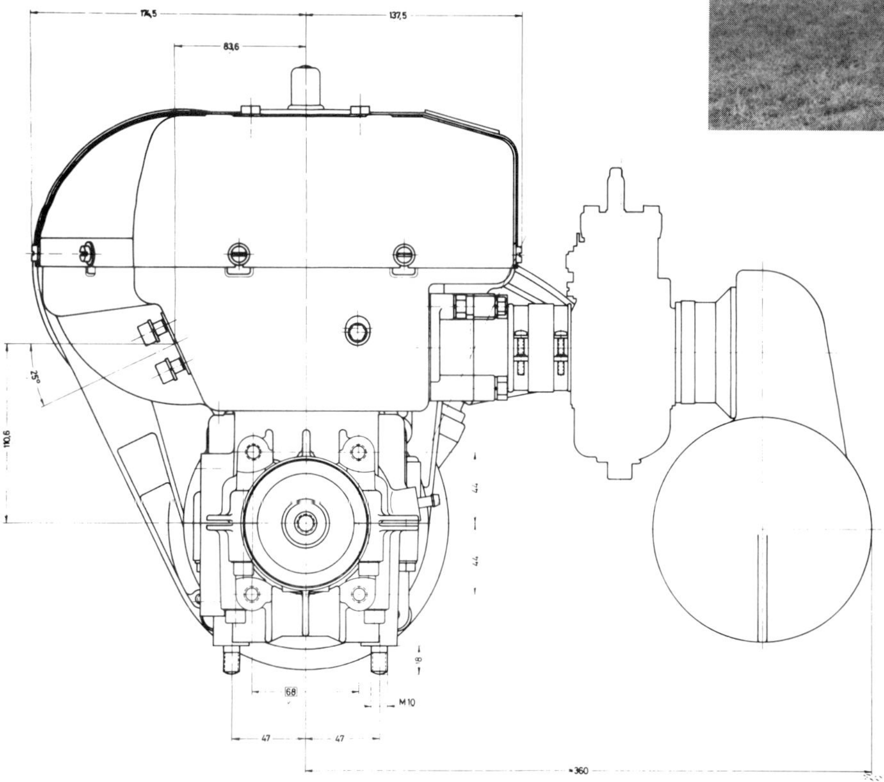

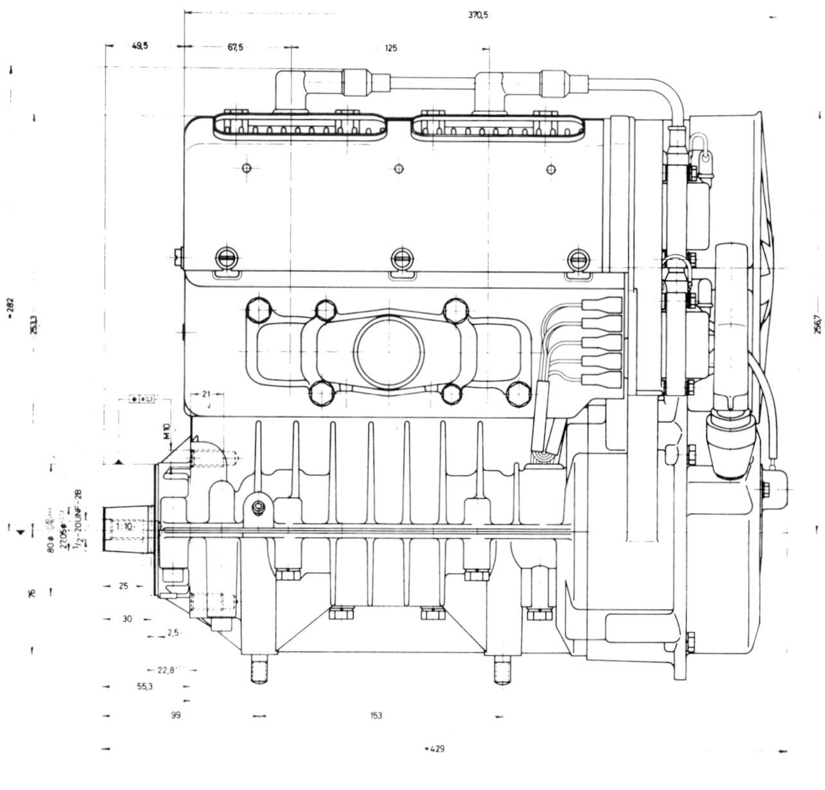

Rotax Engines

Engines from 28 hp to 65 hp, some models free-air cooled, fan-cooled or liquid cooled — with reduction gear of 2.5 to 3.0 to 1. Rotax engines are popular with many ultralight and primary type aircraft owners. The 65 hp liquid cooled Model 532 pictured above was recently installed in a Smith Termite owned by Gerald Olenik, Orwell, Ohio. Note extended engine mount to accommodate this lighter powerplant. Tests were conducted with engine uncowled.

Fuel Systems

The fuel system should be constructed and arranged to ensure a flow of fuel at a rate and pressure adequate for proper engine functioning under any normal operating condition. The system should be reliable under all conditions of flight for which the aircraft is intended. It should be as simple as possible in operation.

The greatest care should be taken to maintain the fuel system in a condition that will eliminate hazards arising from fire and engine failure. Common fuel system engine fires and failures may be attributed to the following:

Fires due to leakage, fumes, insufficient ventilation and drainage, carburetor flooding, overpriming, backfires, improper tank ventilation, or proximity to hot spots such as exhaust manifolds, exhaust discharge, electric sparks or arcing.

Engine failures due to water, dirt, or other foreign matter in the system, vapor lock, air lock, insufficient fuel pressure, improper functioning of fuel system controls, or inaccurate fuel gauge reading, and simply running out of fuel.

17-1 FUEL TANKS

Seams should be located above the fluid level wherever possible.

A common cause of cracking tanks are unsupported flat sections of tanks. Any flat plate areas on metal tanks should be reinforced with angles or channel to support those areas. Unsupported areas are a common cause of tank cracking.

A fuel tank should be able to withstand, without failure, the vibration, inertia, fluid and structural loads that it may be subjected to in operation.

The total usable capacity of the fuel tanks should be sufficient for three hours of flight at normal cruise power, but in any case must be enough for at least one hour of operation at maximum continuous power.

Fuel tanks should be constructed to resist corrosion and should not be susceptible to electrolytic action. This action may be minimized by arranging that dissimilar metals, particularly brass and aluminum are not in close contact with each other. Brass fittings in aluminum tanks create such problems.

Fuel tanks constructed of fiber-glass using polyester resins should be checked for compatibility with the fuel to be used. The use of resins that have been on the shelf for a long or undetermined period should be avoided. It is recommended that tanks be sloshed with a sloshing compound before installation and use.

NOTE: Remember to remove the finger screen prior to sloshing the tank.

It is important that all welding flux be removed before any anti-corrosion processes, sloshing or other coatings are applied.

For tank construction of greatest strength it is usually necessary to have corners with a generous radius and to locate welds on a flat surface removed from the corners. Welding of the tank ends to the body is made easier by forming an expansion bead near the weld.

Where flanges or heavy sumps are put in tanks to take the various connections, they should be large enough and so attached to the tank shell that the stresses induced by the attached connections and piping will not be localized, but will be distributed over a large portion of the tank shell. Reinforcement of the shell at these points of attachment should be made where necessary.

Rivets and welds should be located at points where they will not aggravate the stresses due to vibrations and the sloshing of fuel.

Special attention should be given to riveted joints and seams as sources of possible leakage in tanks of aluminum alloy riveted construction. Soft rivets will usually be found preferable to hard rivets for this reason in tank construction.

Attention should be given to the intersection of two seams where usually special precautions must be taken. In general, except for seams, rivets should be used sparingly, and only where strength or other considerations require their use. Each rivet is a potential source of leakage. Rivet spacing of $3/8$ to $5/8$ inch using rivets $3/32$ to $1/8$ inch in diameter has been found satisfactory in producing leakproof seams in aluminum alloy tanks. It may be necessary to use a double row of rivets to obtain a gasoline-tight joint in larger tanks.

Seams should be located above the fluid level wherever possible.

17-2 FUEL TANK FILLER

Each fuel tank filler cap should be marked with the word "FUEL", the minimum octane permitted, and the usable tank capacity.

All fuel tank caps should be red in color.

The fuel filler neck arrangement should be such that spilled fuel will be prevented from entering the fuel tank compartment or any part of the aircraft other than the tank itself.

Each filler cap should provide a fuel-tight seal for the main filler opening. However, there may be small openings in the fuel tank cap for venting purposes or for the purpose of allowing passage of a fuel gauge through the cap.

17-3 FUEL QUANTITY INDICATOR

Each fuel tank should be provided with a fuel quantity indicator or fuel gauge which will accurately indicate the fuel quantity remaining in the tank.

17-4 UNUSABLE FUEL SUPPLY

Most fuel tanks cannot be drained completely dry but contain a certain residual quantity of fuel called the "unusable fuel". The amount depends upon the shape of the tank and the attitude in which the airplane is flown. It is not safe to rely upon the unusable fuel supply and it is therefore important to determine how much fuel is unusable so that the pilot will have this information available.

The unusable fuel supply for each tank will be established as that level at which the first evidence of malfunctioning occurs under the most adverse fuel feed condition. This should be determined by ground checking the airplane in level and climb attitudes until the engine starts to malfunction due to fuel starvation. At this point each fuel quantity indicator should be adjusted to account for the unusable fuel supply.

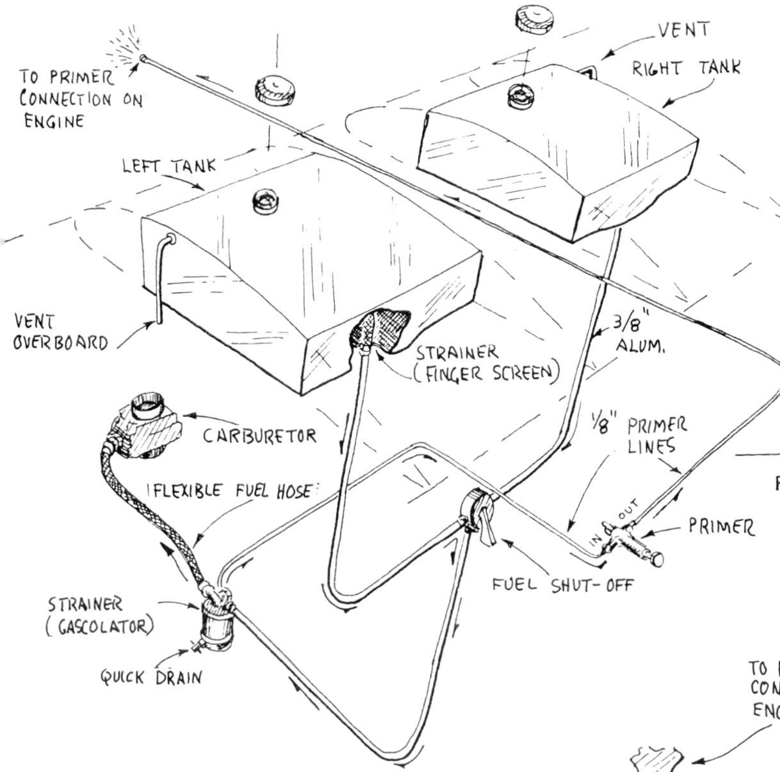

Gravity-Flow Fuel System (High-Wing or Biplane Aircraft)

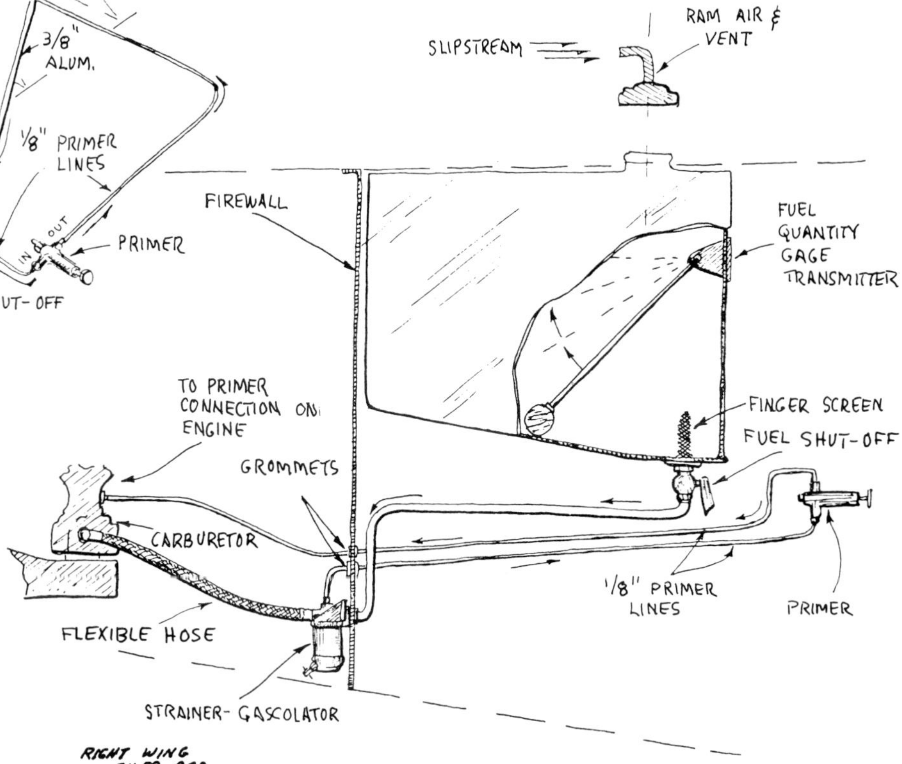

Gravity-Flow System Fuselage Mounted Tank

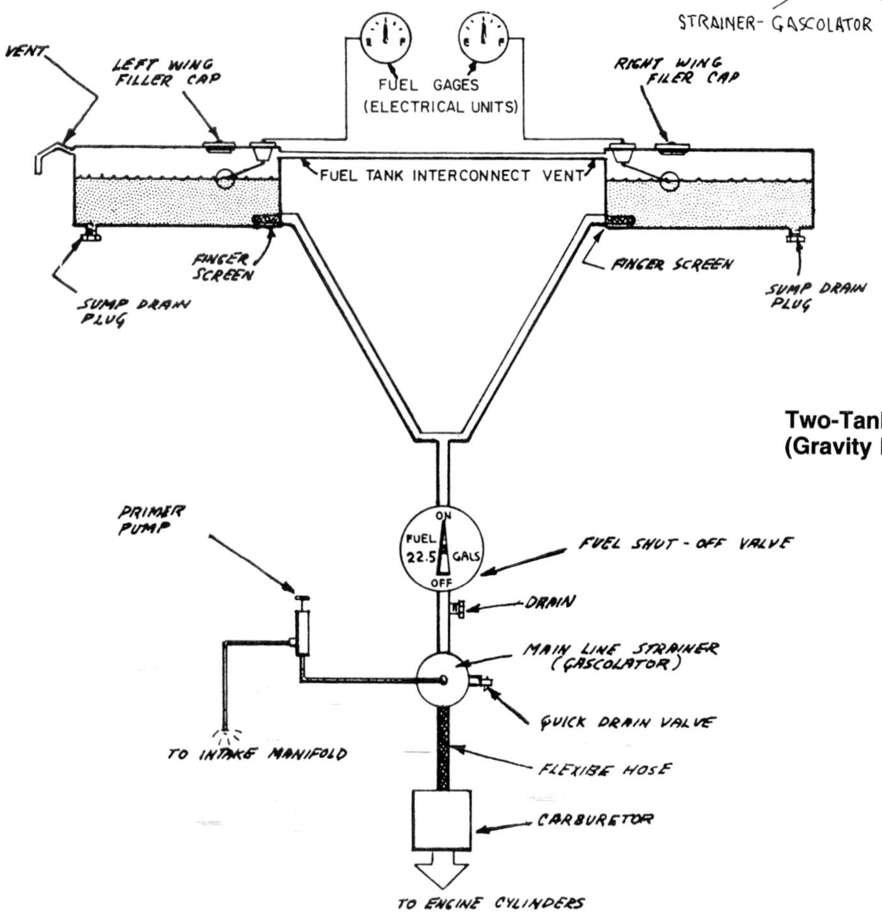

Two-Tank Fuel System Schematic (Gravity Flow System)

Flight checks to determine the amount of unusable fuel supply may be made under carefully controlled conditions but such accuracy is not deemed as necessary.

17-5 FUEL TANK BAFFLES

Baffles or stiffeners should be provided in all tanks over 5 gallon capacity to prevent failure of the tank due to surging of the fuel. This is particularly true of tanks which are either very long or very wide.

Baffle spacing of 12 to 16 inches is suitable.

Baffles should be given careful consideration with regard to their own ability to withstand surging loads and with regard to their attachment to the tank shell to prevent high local stresses from developing.

Each interconnecting passage in baffled tanks should have generous vent and flow openings.

17-6 FUEL TANK EXPANSION SPACE

The filler necks of fuel tanks should be located and installed so as to assure an expansion space of not less than two percent of the tank capacity, unless the tank vent discharges clear of the airplane. In this case no expansion space would be necessary. It should be impossible to fill the expansion space inadvertently with the airplane in the normal ground attitude.

17-7 FUEL TANK VENTS

Each fuel tank should be vented from the top part of the expansion space with the vent located and constructed in a manner that minimizes the possibility of its being obstructed by ice or other foreign matter.

Each vent should be constructed to prevent siphoning of fuel during normal operations.

The venting capacity should be sufficient to allow the rapid relief of excessive differences of pressure between the interior and exterior of the tank. If openings in the filler cap are provided for venting purposes, two or more should be provided for safe operation. Where the float-and-rod type fuel gauge is used, the clearance hole for the rod is considered adequate venting, as this construction has proved satisfactory in service due in part to the vibrating action of the rod.

Airspaces of tanks with interconnected fuel outlets should also be interconnected.

It is undesirable to have undrainable points in any vent line where moisture can accumulate with the airplane in either the ground or level flight attitudes.

No vent may terminate at a location where the discharge of fuel from the vent outlet will constitute a fire hazard or from which fumes may enter personnel compartments.

Each carburetor with vapor elimination connections should have a vent line to lead vapors back to one of the fuel tanks.

For aerobatic category aircraft, excessive loss of fuel during aerobatic maneuvers, including short periods of inverted flight, must be prevented.

It should be impossible for fuel to siphon from the vent when normal flight has been resumed after any aerobatic maneuver.

17-8 FUEL TANK SUMP AND OUTLET

Some tanks have a drainable sump with an effective capacity, in the normal ground and flight attitudes, of at least ½ pint, while others utilize a sediment bowl or gascolator, located in the lowest portion of the fuel system.

The use of sump equipped tanks is not popular with the homebuilder. Due to the use of smaller tanks, reliance is placed, rather, on the use of a gascolator equipped with a quick drain valve.

Each fuel tank outlet should be located so that, in the normal ground attitude, water will drain from all parts of the tank to the sediment bowl or chamber.

It is recommended that all outlets face upwards as side outlets are susceptible to air locking when the upper part of the outlet is uncovered.

17-9 FUEL SYSTEM DRAINS

There should be at least one drain to allow safe drainage of the entire fuel system with the aircraft in its normal ground attitude.

Each drain requires a means to lock it closed.

17-10 FUEL TANK FINGER SCREEN

There must be a fuel strainer or finger screen of 8 to 16 meshes per inch located in the fuel tank outlet. The diameter of the finger screen should be at least equal to that of the fuel tank outlet and its length at least four times the diameter of the outlet. A similar screen permanently mounted in the filler neck of the tank may be considered as equivalent to an outlet screen.

The finger screen should be removable for inspection and cleaning.

17-11 FUEL TANK INSTALLATION CONSIDERATIONS

Past experience has shown that the most reliable type of fuel system is one which feeds the engine from only one tank at a time, regardless of whether a gravity system or a fuel-pump system is used.

Installation should be such that the fuel tank outlets are not uncovered during any maneuvers.

Tanks should preferably feed from a single outlet located so as to be below the fuel level in any normal flight maneuver.

If two outlets are used, care must be taken to make sure there is no possibility of air locking or failure to feed with a low fuel supply.

Fuel tanks should be supported so that tank loads are not concentrated.

Pads should be used, if necessary, to prevent chafing between the tank and its supports. Padding should be nonabsorbent or treated to prevent the absorption of fuel.

Each tank compartment must be ventilated and drained to prevent the accumulation of flammable fluids or vapors. Each compartment adjacent to a tank that is an integral part of the aircraft structure should also be ventilated and drained.

Ventilation is considered adequate from a safety standpoint when a tank of less than 25 gallons is located underneath or ahead of the instrument panel, a typical installation in a small airplane.

No fuel tank may be on the engine side of the firewall.

There should be at least one-half inch of clearance between the fuel tank and the firewall.

17-12 FUEL PUMP TYPE SYSTEM

The fuel-pump system uses an engine-driven pump instead of gravity to supply fuel pressure. Since the fuel pressure derived from a fuel pump is not dependent upon the height of the fuel tanks above the engine, this system is especially practical in low wing aircraft where the fuel tanks are located in the wings below the level of the carburetor.

The fuel pump should be directly driven by the engine. Diaphragm-type fuel pumps should be considered for hazards resulting from ruptured diaphragms. A small gravity tank, fed by the diaphragm fuel pump, feeding the carburetor is recommended.

To provide a back-up system for the engine driven pump an emergency hand, or separate power-driven, pump of adequate capacity should be installed and be available for immediate use in case of pump failure, particularly at take-off. Frequently an electrical fuel-boost pump is installed in the system with other components or submerged in the tanks themselves.

A fuel pump system should be arranged to allow a fuel pump to draw fuel from one tank at one time. When two tanks are used at the same time, with a pressure fuel system, it has been found difficult to make sure that fuel will feed evenly from each tank and so the tank which is emptied first may air lock the system.

17-13 GRAVITY-FLOW FUEL SYSTEM

A gravity feed system may supply fuel to an engine from not more than one tank at a time, unless the tank airspaces are interconnected so that the interconnected tanks will feed equally.

When two tanks are used at the same time with a gravity system, without interconnecting their vents, a difference in pressure at the vents can cause fuel flow between tanks. Then, either early fuel exhaustion from one tank or overflow from the other causes a fire hazard.

A gravity feed system with interconnected tank outlets should be so constructed as to make it impossible for enough fuel to flow between the tanks to cause an overflow of fuel from any tank vent.

A gravity system should provide fuel at a rate equal to one and one half (1½) times the actual take-off fuel consumption of the engine.

17-14 FUEL SYSTEM STRAINER/GASCOLATOR

The fuel strainer or gascolator is located at the lowest point in the fuel system and is used tc trap any foreign particles or moisture that might accumulate in the system. The location of the gascolator at this point and the restrictive effect of the filter element, permits the water to be separated from the fuel and collected in a sump bowl at the base of the strainer assembly.

There must be a fuel strainer between the fuel tank outlet and carburetor inlet (or engine driven fuel pump, if there is one installed).

Each such strainer or filter must be accessible for drainage or cleaning and should discharge clear of the aircraft. It should permit the complete drainage of the tank and the fuel system when the aircraft is in its normal ground attitude.

17-15 CRASH PROTECTION OF FUEL SYSTEM COMPONENTS

Fuel system components, especially the gascolator, should be so installed as to be protected from damage, which could cause the release of fuel, as a result of a collapsed gear or a wheels-up landing. Information on safety fuel cells are available from sources known to EAA Headquarters and can be obtained by contacting them. There have been some very bad accidents that could have been prevented by fuel cells like the race driver's use.

17-16 FUEL LINES AND FITTINGS

Fuel lines and fittings require considerable thought and attention:

Each joint is a possible source of leakage.

Each bend is a possible blockage source.

Each rise in the fuel line is a possible source of vapor lock.

Each low spot is a possible source of water collection and subsequent freezing.

Each length of unsupported tubing is susceptible to vibrational fatigue.

Each hole through which the fuel line passes is a possible source of chafing.

Lines and fittings which are under pressure, or which attach directly to the engine, or which are attached between components that may move in relation to each other, should be flexible and fire resistant lines with approved type fire resistant end fittings. Other lines need not be flexible but should be of fire resistant materials.

Fuel lines, except flexible portions, should be supported by soft blocks and clamps. Friction tapes and rawhide lashings tend to collect grit and cause wear due to movement

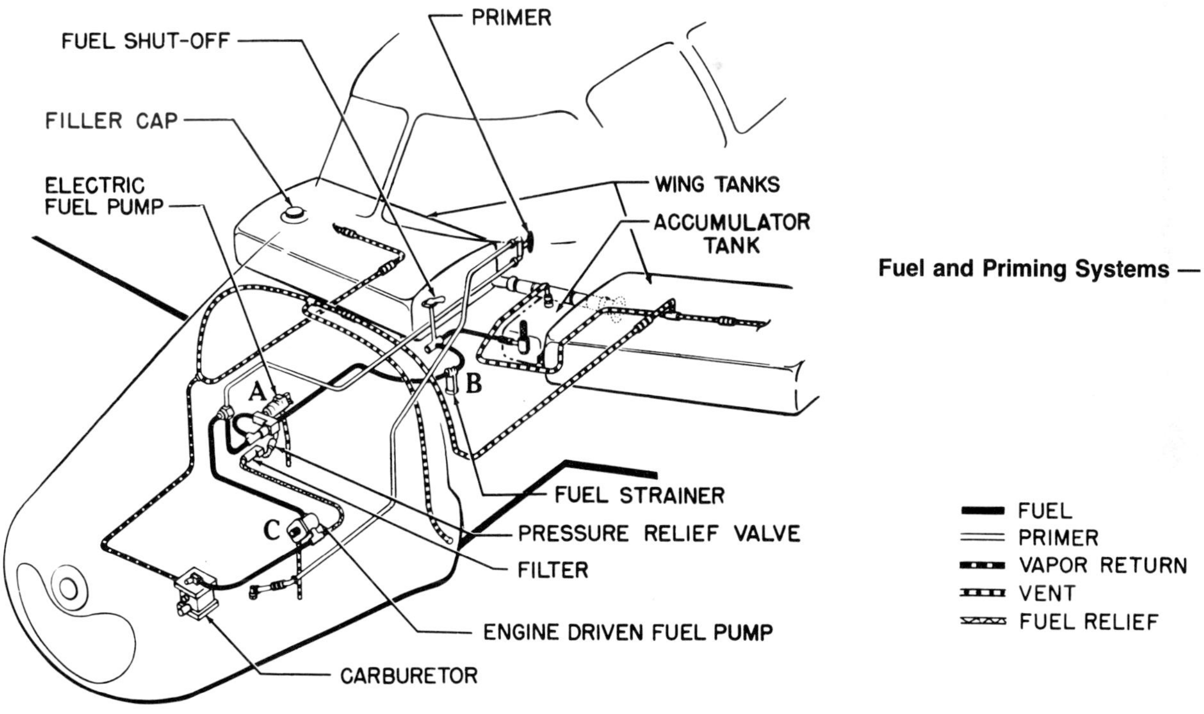

Fuel and Priming Systems — Navion

Pitts Specials are one of the finest aerobatic aircraft ever to take to the skies. Jim Norton, Kent City, MI competes at IAC competitions at Fond du Lac, WI with this Pitts S1D.

which causes the lines to work under the lashings.

Although the minimum recommended size for fuel feed lines is tubing of 3⁄8" OD × .032" wall, the fuel line required may be determined by the size of the fitting at the carburetor.

Lines up to 18" in length usually do not require supports. Short solid tubing with flexible connections at each end need not be supported.

Bends in the fuel line of small radius, vertical humps, or restrictions which might help cause vapor or airlocking in the lines should be avoided wherever possible.

The radius of a tube bend should not be less than three times the outside diameter of the tubing.

17-17 FUEL SELECTOR VALVES

There should be a means to allow the pilot to rapidly shut off the fuel to each engine individually while in flight or on the ground.

Fuel valves should be quick-acting, designed for the purpose, and not susceptible to external leakage. They should be as easy to open as to close. In the "on" and "off" positions, they should have positive stops or "feel".

All fuel valves should be supported in such a way that flight or operating loads should put no strain on the fuel lines connected to the valves.

Markings to show the direction of operation and function at each stop are essential. It is important that the valves be so installed that the effect of gravity and vibration will be to turn the valve on rather than off.

Fuel valves should be mounted in such a position that there will be little chance of the pilot's hand, arm, leg or clothing striking the valve and changing its setting.

The valve handle must clearly indicate the position of the valve and it should be so arranged that the handle cannot be mounted in an incorrect position.

If possible, the fuel valve handle should not be mounted near the carburetor heat, cabin heat, or any other frequently operated control unless protected in some way to avoid accidental operation.

17-18 FUEL FLOW TEST

The fuel system must be capable of providing fuel at a rate and pressure sufficient for proper carburetor operation in the attitude that is most critical with respect to fuel feed and quantity of unusable fuel. These conditions may be simulated in a suitable mockup or conducted in the aircraft.

To check the fuel flow using the aircraft will require that the aircraft be blocked up with the thrust line at the best angle of climb with take-off power. This angle is usually greater than the three point landing attitude and should be estimated accordingly.

The flow should be measured at the carburetor inlet and restricted at that point to the minimum operating pressure for a satisfactory take-off mixture as recommended by the carburetor and engine manufacturers.

A satisfactory way to check the inlet fuel pressure is to disconnect the feed line at the carburetor inlet and, at this level, connect it to a pressure gage. If the gage reads higher than ½ psi, the pressure is normally adequate. A reducing nipple used at the end of the feed line will raise the pressure, if necessary. If a reducing nipple is used, the flow should be measured through the nipple. The rest of the fuel system should be left intact as it would be in service.

Before beginning the test, the fuel system should be completely drained. If there is more than one tank, the system should be set to feed from one only.

Each tank should be tested separately.

Fuel should be added to the tank being tested slowly until a steady flow is created at the carburetor end of the feed line. Steady flow should be established when approximately the unusable fuel supply has been added. Systems requiring an excessive amount of fuel to produce steady flow are unsatisfactory since they tend to airlock.

After the steady flow has been established an additional gallon of fuel should be added.

The time in seconds required for at least 1 gallon of fuel to flow from the feed line should then be obtained.

The time required for 1 gallon to flow should not be longer than the following: Provide a factor of safety with 150% of actual takeoff fuel consumption of the engine.

NOTE: Specific fuel consumption, etc. can be obtained from the engine manufacturer's maintenance manual or the FAA.

As an example, for an engine of 100 hp which consumes 60 lbs./hr. or 10 gals./hr. use a factor of safety and multiply the 10 gals./hr. × 1.5. This will leave you with an expected fuel flow of 15 gals./hr. which is the equivalent of 1 gallon every 4 minutes. Therefore, a gallon of fuel should not take longer than 4 minutes to flow through the system. However, most float carburetors require at least ½" PSI pressure to operate properly. For this reason, the hose should be disconnected at the carburetor and lifted 20" higher than the inlet point to provide at least ½" of pressure before checking for proper fuel flow. In addition, the aircraft should be placed in the highest climb attitude it will normally be expected to climb at by placing the wheels on blocks, tying the tail down, etc. Use extreme care in this operation to avoid damage to the aircraft or injury to yourself.

17-19 FUEL TANK TESTS

Each fuel tank must be able to withstand the following pressures without failure or leakage:

For each conventional metal tank and nonmetallic tank with walls not supported by the aircraft structure, a pressure of 3 PSI is used. It is best to use water pressure, not air pressure; in the event of a rupture, no injury will result. A column of water 98" high (any diameter) will provide a pressure of 3.5 PSI on a fuel tank. Drain **all** the water from the tank after the test.

For each integral tank, the pressure estimated to be developed during the maximum limit acceleration of the aircraft with a full tank.

17-20 PRIMER

A small mounted hand pump is used to pump fuel directly into the engine cylinders or intake manifolds to aid in engine starts. It is particularly useful for cold weather starts. It should be installed where it is easily accessible to the pilot.

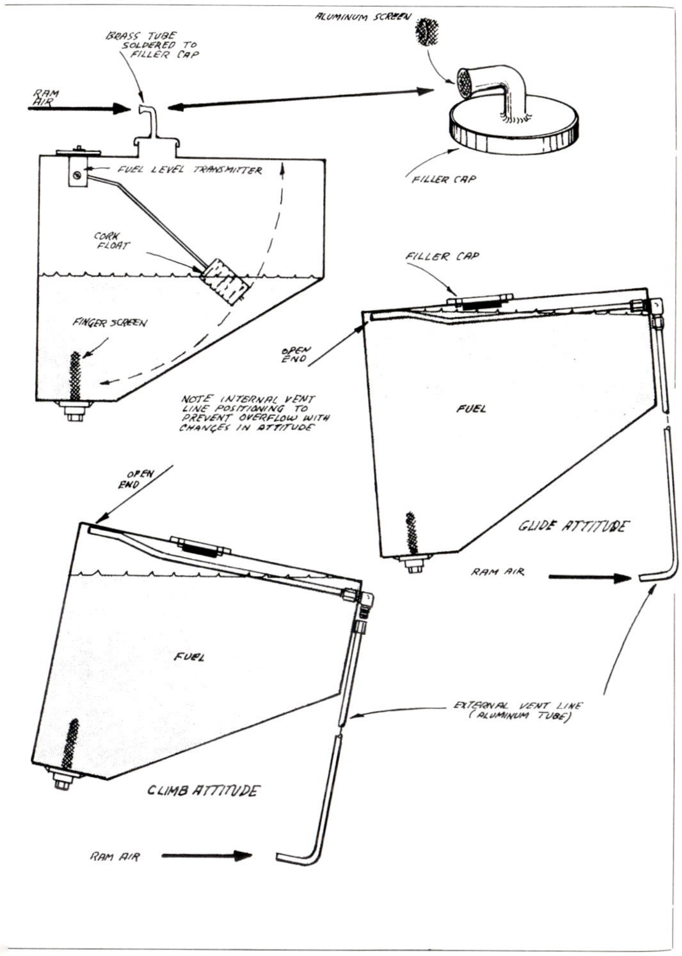

Gravity Flow Fuel System
(No Fuel Management Problems Here!)

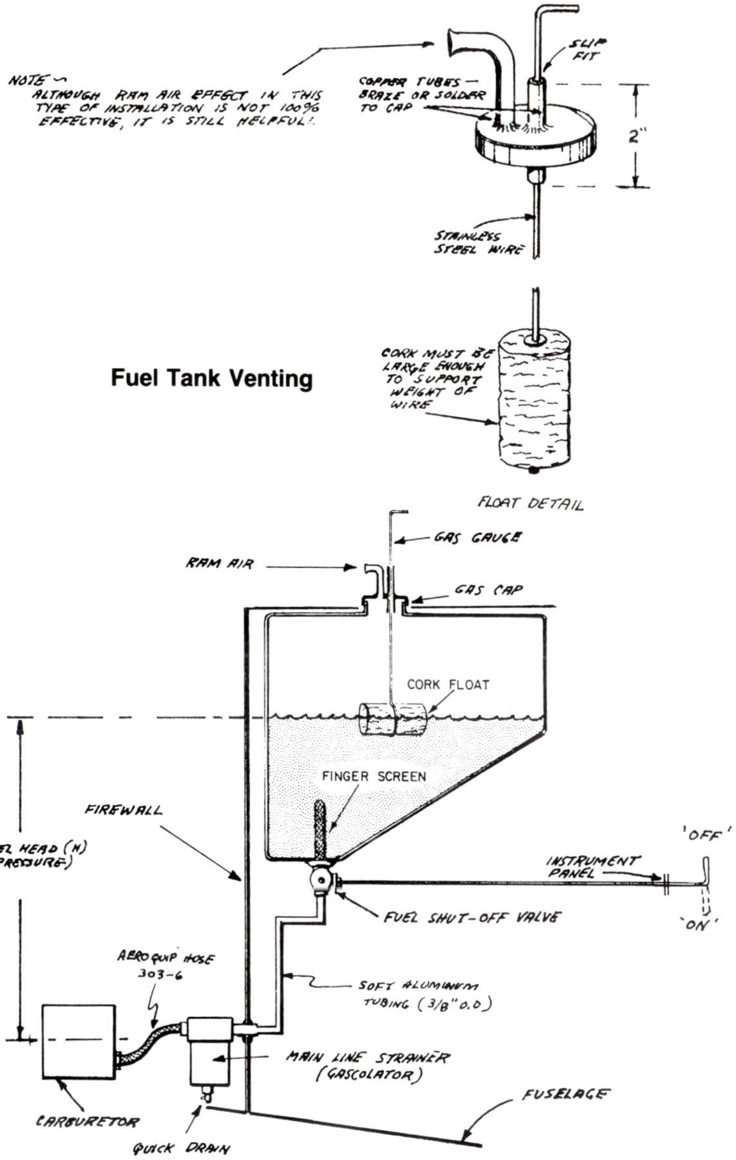

Fuel Tank Venting

Steve Wittman, Oshkosh, WI has been building homebuilt airplanes since the late 20s. His first homebuilt racer was the Chief Oshkosh, a 90 hp Cirrus powered speedster that became one of the winningest racers of all time. Shown here is Steve's equally famous Formula 1 racer Bonzo which he completed in 1948. He dominated the pylons with it throughout the late 40s and 50s. At age 87 Steve still flies. In 1968 Steve won all three categories of the Lowers-Baker-Falk 500 mile efficiency race.

Oil System

18-1 OIL TANK CAPACITY AND EXPANSION SPACE

Each 4 cycle reciprocating type engine must have an independent oil system that can supply it with an appropriate quantity of oil at a temperature not above that safe for continuous operation. Other engines may require lubrication systems unique to their own types.

However the usable oil supply may not be less than the product of the endurance of the airplane under critical operating conditions and the maximum oil consumption of the engine under the same conditions, plus a suitable margin to ensure adequate circulation and cooling.

Each oil tank should have an expansion space of not less than 10 percent of the tank capacity. It should be impossible to fill the expansion space inadvertantly with the airplane in the normal ground attitude.

18-2 FILLER CONNECTION

Each oil tank filler cover must be marked with the word "OIL" and the tank capacity. The filler cap is to be **yellow** in color.

18-3 VENTS

Each oil tank is to be vented to the engine crankcase from the top part of the expansion space so that the vent connection is not covered by oil under any normal flight condition.

Oil tank vents should be arranged so that condensed water vapor that might freeze and obstruct the line cannot accumulate at any point.

For acrobatic category airplanes, there should be means to prevent hazardous loss of oil during acrobatic maneuvers, including short periods of inverted flight.

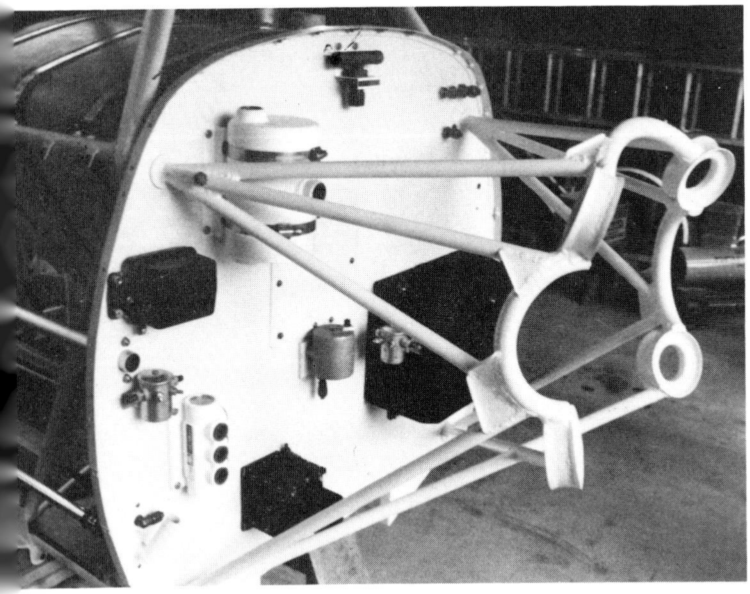

Dynafocal mount for Lycoming 0-360 A4A. Mounted on the firewall of this partially completed Skybolt is the oil valve (unit with three openings), oil separator (large round tank), and the breather T which are all part of the inverted oil system installed on most aerobatic type aircraft.

18-4 OUTLETS

No oil tank outlet should be enclosed or covered by any screen or guard that might reduce the flow of oil. No oil tank outlet diameter should be less than the diameter of the engine oil pump inlet.

18-5 OIL TANK TESTS

Each custom-made oil tank should be pressure tested to an applied pressure of 5 PSI. A column of water, any diameter, 140" high will provide this head of water pressure.

18-6 OIL LINES

Oil lines must be of an acceptable type and must accommodate a flow of oil at a rate and pressure adequate for proper functioning under any normal operating condition.

18-7 BREATHER LINES

Breather lines must be arranged so that condensed water vapor that might freeze and obstruct the line cannot accumulate at any point. The breather discharge must not constitute a fire hazard if foaming occurs, or cause emitted oil to strike the pilot's windshield. The breather should not discharge into the engine air induction system. Acrobatic category airplanes must not experience excessive loss of oil from the breather during acrobatic maneuvers and during short periods of inverted flight.

18-8 OIL STRAINER OR FILTER

Each oil strainer or filter in the powerplant installation must be constructed and installed so that oil will flow at the normal rate through the rest of the system with the strainer or filter element completely blocked.

18-9 OIL SYSTEM DRAINS

There must be at least one accessible drain that allows safe drainage of the entire oil system. The drain should have a manual or automatic means for positive locking in the closed position.

18-10 OIL RADIATORS

Each oil radiator and its supporting structures must be able to withstand the vibration, inertia, and oil pressure loads to which it would be subjected in operation.

Hydraulic Systems

19-1

Each hydraulic system and its elements must withstand, without yielding, the structural loads expected in addition to hydraulic loads.

19-2

Each system should be capable of withstanding a load at least 1.5 times the maximum operating pressure of that system.

19-3

Retractable landing gears should be thoroughly checked with the aircraft blocked up on its jackpoints located far enough outboard on the wing to simulate wing bending as will occur in flight. The gear should then be retracted and recycled to insure proper operation and clearance of all components.

Hydraulic toe brakes as installed on the Pober Pixie.

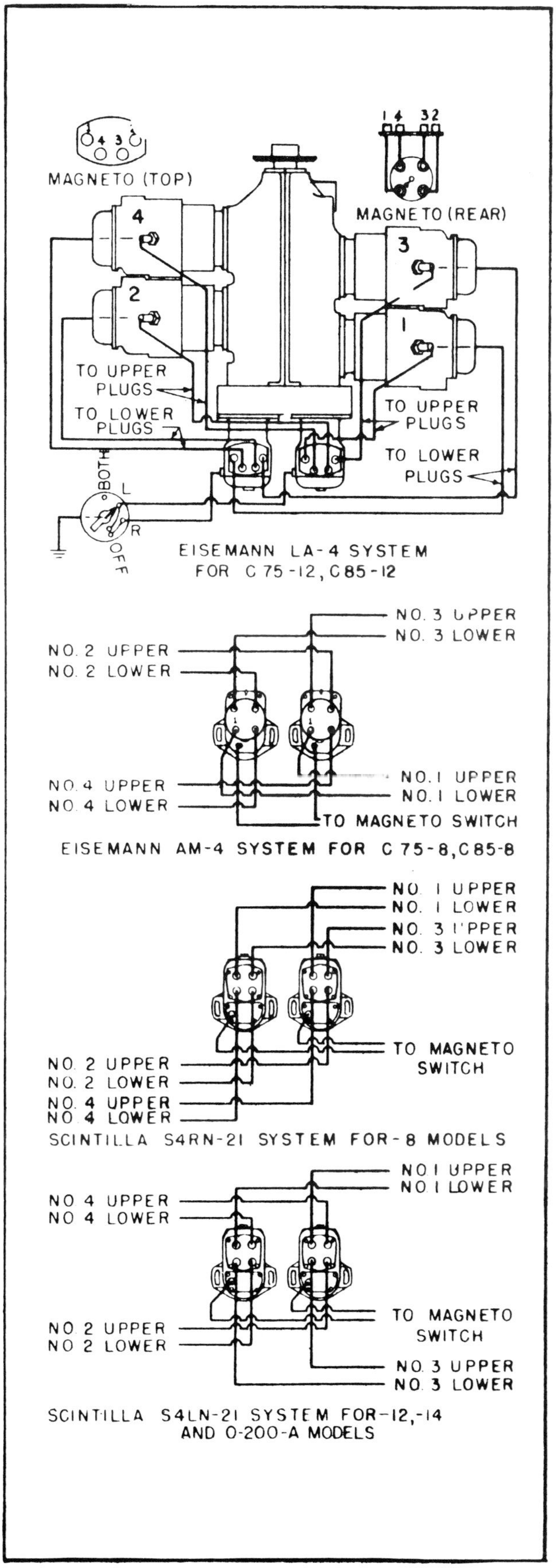

IGNITION DIAGRAM
0-200 CONTINENTAL

Control Systems

20-1 PRIMARY FLIGHT CONTROLS

Each control must operate easily, smoothly, and positively enough to allow proper performance of its functions.

Controls should be arranged and identified to provide for convenience in operation and to prevent the possibility of confusion and subsequent inadvertent operation.

Primary flight controls are those used by the pilot for the immediate control of pitch, roll and yaw.

20-2 COCKPIT CONTROLS — THEIR MOVEMENT AND EFFECT

Cockpit controls must be designed and installed so they operate as follows:

Aileron . . . Right (clockwise) movement of control for right wing down.

Elevator . . . Rearward movement of control for nose up.

Rudder . . . Right pedal forward for nose right.

20-3 INSTALLATION CONSIDERATIONS

An operations test should prove that when the controls are operated from the pilot compartment with the system loaded slightly at the control surfaces, the system is free from jamming, excessive friction, and excessive deflection.

20-4 STRUCTURAL RIGIDITY

Special care should be taken to provide sufficient strength to carry concentrated loads from aileron and flap hinges into the main structure.

It is essential, too, that control surfaces, when subjected to operational conditions, have sufficient torsional rigidity and indicate no signs of excessive deflection or permanent set.

No specific limits of permissible maximum deflection of the surface alone are offered, since these may vary widely with the type, size and construction of the control surface.

20-5 COCKPIT CONTROLS

The predominating type of cockpit controls is the stick and pedal system. Wheel control is sometimes preferred for high performance aircraft with restricted cockpit size.

20-6 POSITIONING OF CONTROLS

In the installation of any control, consideration should be given to its relative importance and to its convenient placement for the usual sequence of operations. Thus for landing, it is desirable that flap control and brakes be operable without changing hands on the wheel or stick. Likewise, secondary controls should be so located that the possibility of accidental or mistaken operation is remote.

20-7 CONTROL SENSITIVITY

It is suggested that the total travel at the top of the conventional stick should be at least 9 inches or more in both planes to avoid undue sensitivity of the elevators and provide sufficient leverage on the ailerons.

When a wheel control is used the angular motion should be not less than 60 degrees either side of neutral.

Rudder pedals should have at least 2 inches of travel either way.

20-8 CONTROL SYSTEM FRICTION

Excessive friction should be especially avoided in the control system. This may sometimes dictate the use of special anti-friction bearings at heavily loaded pivots.

Excessive cable tension usually increases system friction.

20-9 CONTROL INTERFERENCE

Proper precautions should be taken with respect to control systems to eliminate the possibility of jamming, interference from baggage, passengers or loose objects, and chafing or slapping of cables against parts of the aircraft.

All pulleys should be provided with satisfactory guards.

A control column or stick located between a pilot and a passenger should not be used unless a throw-over type of wheel control is incorporated.

Control columns should be protected by a flexible boot as a guard against possible fouling in flight by foreign objects.

20-10 CONTROL SURFACE CLEARANCES

Clearances between any movable control surface and stationary structure should be sufficient to insure the absense of interference.

At the control surfaces themselves ample clearances should be left beyond the normal deflections, possibly 5 degrees, to prevent interferences and possible damage when the surface is slammed over by wind on the ground.

20-11 NOSE WHEEL CONTROLS

It is essential when a nose wheel steering system is interconnected with the flight controls that care be taken to prevent excessive loads from the nose wheel overstressing the flight control system. This objective may be attained by springs, a weak link, or equivalent means incorporated in the nose wheel portion of the control system.

20-12 CABLE TENSION CHANGES

The movements of horns, cables and other components with respect to each other should be such that there is no excessive change in system tension throughout the range. Elevators mounted on adjustable stabilizers, in particular should be checked for this possibility.

Pulley guards should be close fitting to prevent jamming from slack cables since wide temperature variations may cause rigging loads to vary appreciably.

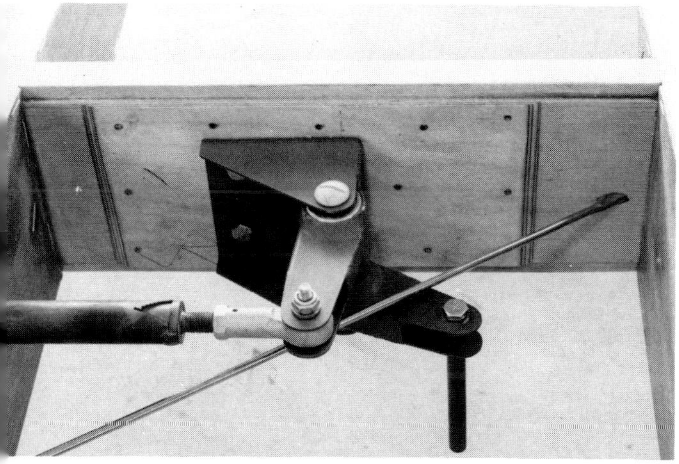

Top — Aileron bell crank assembly showing adjustable push-pull tube and rod end bearings — wing to aileron. Center — Bell crank temporarily fitted into position for final alignment and to check clearances. Bottom — Tail group of Acro Sport II is large and adds to the aircrafts fine flying qualities. Elevator contains both trim and servo tabs (not installed here).

20-13 AERODYNAMIC BALANCE

When using extreme values of differential motion in the aileron control system or a high degree of aerodynamic balance of the ailerons, the friction in the system must be kept low; otherwise the ailerons will not return to neutral and the lateral stability characteristics will be adversely affected. This is particularly true when the ailerons are depressed as part of a flap system, in which case there may even be definite over-balance effects.

20-14 ADJUSTABLE STABILIZERS

Adjustable stabilizer controls should be free from "creeping" tendencies. When adjustment is secured by means of a screw or worm, the lead angle should not exceed 4 degrees unless additional friction, a detent, or equivalent means is used. In general, some form of irreversible mechanism should be incorporated in the system, particularly if the stabilizer is hinged near the trailing edge.

20-15 ELEVATORS AND RUDDERS

When separate elevators are installed they should be rigidly inter-connected.

When dihedral is incorporated in the horizontal tail the universal connection between the elevator sections should be rugged and free from play.

External wire bracing on tails is subject to vibration and the design of the wire assembly and end connections should be such as to withstand this condition.

Leading edges and struts should have adequate strength to withstand loads if handles or grips are not provided.

Movable tail surfaces should be so installed that there is no interference between the surface or their bracing when anyone is held in its extreme position and any other is operated through its full angular movement.

20-16 CABLES, PULLEYS & FAIRLEADS

Each cable, cable fitting, turnbuckle, splice and pulley used should be of an approved type or aircraft quality hardware.

Control cables should be of the ⅛" diameter 7 × 19 extra-flexible type.

Cable splices should be made by use of Nico-Press sleeve and thimble or of approved type swaged terminals. Cable sizes may be governed by other considerations as well as by strength requirements, therefore, 3/32 inch diameter 7 × 19 extra flexible cable might be considered adequate for ultra-light aircraft.

Turnbuckles should be located so as to be accessible for adjustment and preferably not in the center of long unsupported spans where they can slap around too freely.

Fairleads of non-metallic material, such as the phenolic plastic compounds should be used to prevent cables, chains and links from chafing or slapping against parts of the aircraft, but should not be used to replace pulleys as a direction-changing means. However, where the cable load is small, and the location is open to each visual inspection, direction changes (through fairleads) not exceeding 3 degrees are satisfactory in primary control systems. A somewhat greater value may be in secondary control systems.

Pulleys that carry a 180 degree cable bend should be carefully mounted so that there is no danger of their leaning over and binding under load. They should also line up accurately with the plane of the cables . . . or the flanges will wear out quickly.

All pulleys should be provided with annular guards so located that a slack cable cannot get off the pulley in any manner and jam it.

20-17 CONTROL SURFACE HORNS

Control surface horns are generally constructed of metal. They should be attached to the control surface spars near a hinge to avoid bending the spar, and in such a manner as to distribute the loads into the spar.

Although many high performance aircraft continue to use external control horns, it is believed that internal horns are a worthwhile attempt to reduce drag.

20-18 HINGES

Hinges should be as friction-free as possible. Replaceable bushings and sleeves of suitable material should be provided for bearing surfaces wherever practical to do so.

Bearings should be arranged so that they can be readily inspected. Provisions for lubrication should be made if self-lubricated or sealed bearings are not used.

Clevis pins may be used as hinge pins provided that they are made of suitable material and are properly locked.

The effects of deflection of the surfaces, such as in bending, should be allowed for, particularly with respect to misalignment of the hinges. This may also influence spacing of the hinges.

Sufficient restraint should be provided in one or more brackets to withstand forces parallel to the hinge center line. Rudders, for instance, may be subjected to high vertical accelerations in ground operation.

Hinges welded to elevator torque tubes or similar components may prove difficult to align unless kept reasonably short and welded in place in accurate jigs.

Piano-type hinges are acceptable with certain restrictions. In general, the attachment should be made with some means other than wood screws and this attachment should be as close as possible to the hinge line to reduce flexibility. Piano hinges should not be used at points of high loading, such as exist at control horns, unless the reaction is satisfactorily distributed. Due to possible binding and to the difficulty in inspecting or replacing a worn hinge wire, it is better to use several short lengths than one long piano hinge.

20-19 CONTROL SURFACE STOPS

Stops are advisable for all control surfaces, particularly adjustable stabilizers and elevator trailing edge tab systems. For these, the stops should be positioned so as to limit the travel to the required range. In general, stops are advisable for all surfaces in order to avoid interferences and possible damage to the parts concerned, particularly in the case of large surfaces where the deflections may permit the surface to exceed the design range of travel.

Series of holes in attach fitting permit easy rudder pedal adjustment.

Each stop must be able to withstand the loads imposed on it in normal use.

Whenever the location of stops within the control system is not specified, they should preferably be located close to the operating force in order to avoid a "spongy" control.

Ideally, additional stops also should be installed at the control surfaces.

20-20 CONTROL SYSTEM LOCKS

When a device is provided for locking a control surface while the aircraft is on the ground or water, the locking device should be so installed as to positively prevent taxiing or taking off, either intentionally or inadvertently, while the lock is engaged.

Some means should be provided to preclude the possibility of the lock becoming engaged during flights.

20-21 FLAP CONTROLS

The flap operating mechanism should be such as to prevent sudden inadvertent or automatic opening of the flaps at speeds above the design speed for the extended flap conditions.

Means should be provided to retain flaps in their fully retracted position and to indicate such position to the pilot.

Undesirable flight characteristics, such as loss of lift and consequent settling, may result from too rapid operation of flaps which give appreciable lift. When the prime function of the flap is to act as a brake, however, slow operation is not as important.

The control and means for the interconnection of the flaps should insure that the flaps on both sides function simultaneously.

20-22 TRIM SYSTEMS

Position Indicator - When adjustable elevator tabs are used for the purpose of trimming the aircraft, a tab position indicator should be installed and means provided for indicating to the pilot a range of adjustment suitable for safe takeoff and the directions of motion of the control for nose-up and nose-down motions of the aircraft.

Control Tab Reversibility - Tab controls should be irreversible unless the tab is statically balanced about its hinge line and has no unsafe flutter characteristics.

Tab Wear and Vibration - In addition to the air loads, consideration should be given in the design to the lapping effect of dust and grease on fine threads, deflections of the tab due to the small effective arm of the horn or equivalent member, and vibration common to the trailing edge portion of most movable surfaces.

Trim Tab Sensitivity - It is advisable to avoid a tab cockpit control with small travel because of the resulting abrupt action of the tab.

Operation of Tabs - Proper precautions should be taken against the possibility of inadvertent or abrupt tab operation and operation in the wrong direction.

If trailing-edge tabs are installed, care should be taken in proportioning areas and relative movements so that the main surface is not aerodynamically overbalanced at any time.

Minimum deflections and play are of primary importance in the installation of tabs. Strength of the surface and anchorages should be sufficient to prevent damage or misalignment from handling, especially thin sheet tabs which are set by bending to the proper position.

Electrical Equipment

21-1 ELECTRICAL SYSTEM

Electrical equipment, if installed, should be free from hazards in themselves, in their method of operation and in their effects on other parts of the aircraft. Items of equipment which are essential to the safe operation of the aircraft should be designed and installed to insure that they will perform their intended functions reliably during all reasonably foreseeable operating conditions.

The following miscellaneous equipment is required if an electrical system is installed:

A master switch arrangement.

An adequate source of electrical energy.

Electrical protective devices.

There should be an ammeter installed in the aircraft for proper monitoring of the aircraft's electrical power.

21-2 STORAGE BATTERY

The capacity of storage batteries should be sufficient to supply all connected loads in probable combinations and for probable durations. If corrosive electrolyte can be spilled during flight or during servicing, means should be provided to prevent hazardous corrosive effects on adjacent structure or equipment. The battery should be accessible for service and inspection on the ground.

21-3 MASTER SWITCH

If electrical equipment is installed, a master switch should be provided which will disconnect the storage battery from all loads at a point adjacent to the battery.

21-4 CIRCUIT PROTECTIVE DEVICES

Protective devices (fuses or circuit breakers) should be installed in the circuits to all electrical equipment. Protective devices in circuits essential to safety in flight should be accessible and identified so that they may be reset (or replaced, in the case of fuses) during flight. Sufficient spare fuses should be provided for this purpose.

21-5 INSTRUMENT LIGHTS

Instrument lights, when required, should provide sufficient illumination to make all instruments and controls easily readable. The lights should be installed in such manner that direct rays or bright reflections are shielded from the pilot's eyes.

A cabin dome light is not an instrument light.

21-6 ANTICOLLISION LIGHT SYSTEM

If certification for night operation is requested, the airplane must have an approved anticollision light system, either TSO'd or documented to meet TSO specifications.

21-7 POSITION LIGHT SYSTEM INSTALLATION

Each part of each position light system must meet the visual requirements required of such systems and be TSO'd or meet the requirements of the TSO.

21-8 FORWARD POSITION LIGHTS

Forward position lights must consist of a **red** and a **green** light spaced laterally as far apart as practicable and installed forward on the airplane so that, with the airplane in the normal flying position, the **red light** is on the **left side** and the **green light** is on the **right side**.

21-9 REAR POSITION LIGHT

The rear position light must be a **white light** mounted as far aft as practicable.

21-10 POSITION LIGHT CIRCUIT

The two forward position lights and the rear position light should make a single circuit.

21-11 LIGHT COVERS AND COLOR FILTERS

Each light cover or color filter must be at least flame resistant and may not change color or shape or lose any appreciable light transmission during normal use.

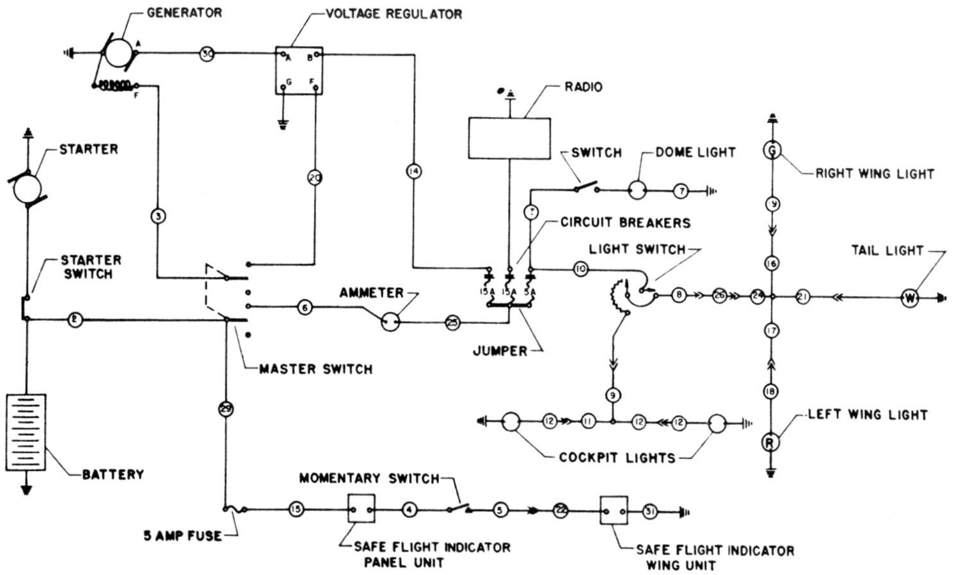

ELECTRICAL SYSTEM DIAGRAM
MODEL PA-20

SINGLE ENGINE GENERATOR AND STARTER WIRING DIAGRAM

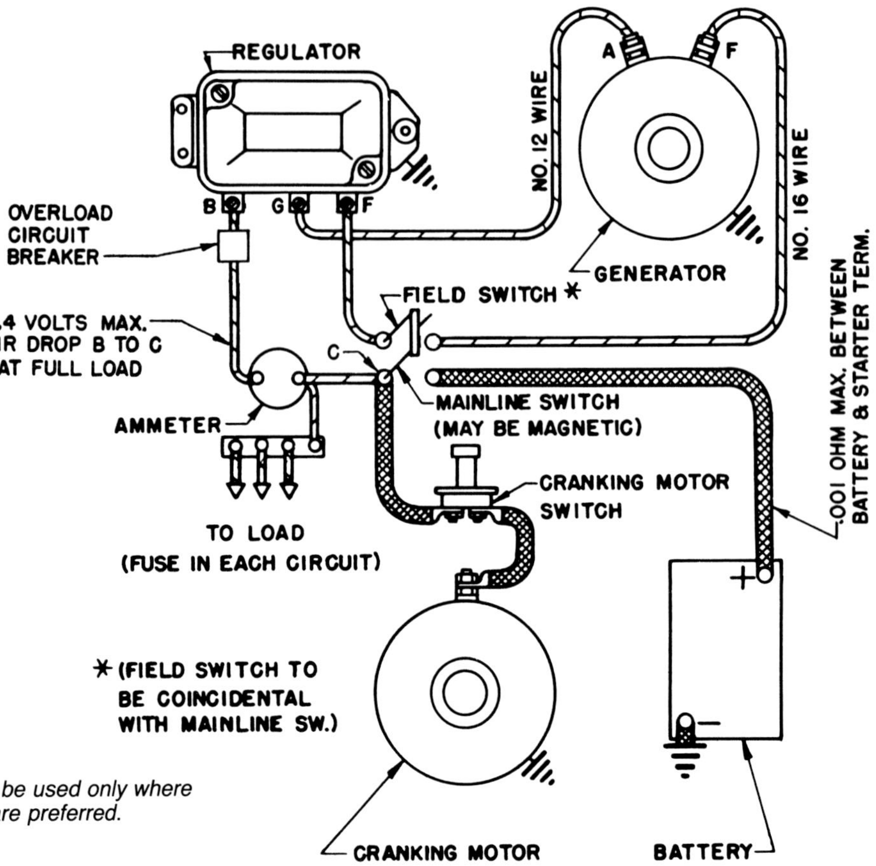

Wire sizes shown are minimum and should be used only where lead lengths are short. Heavier wire sizes are preferred.

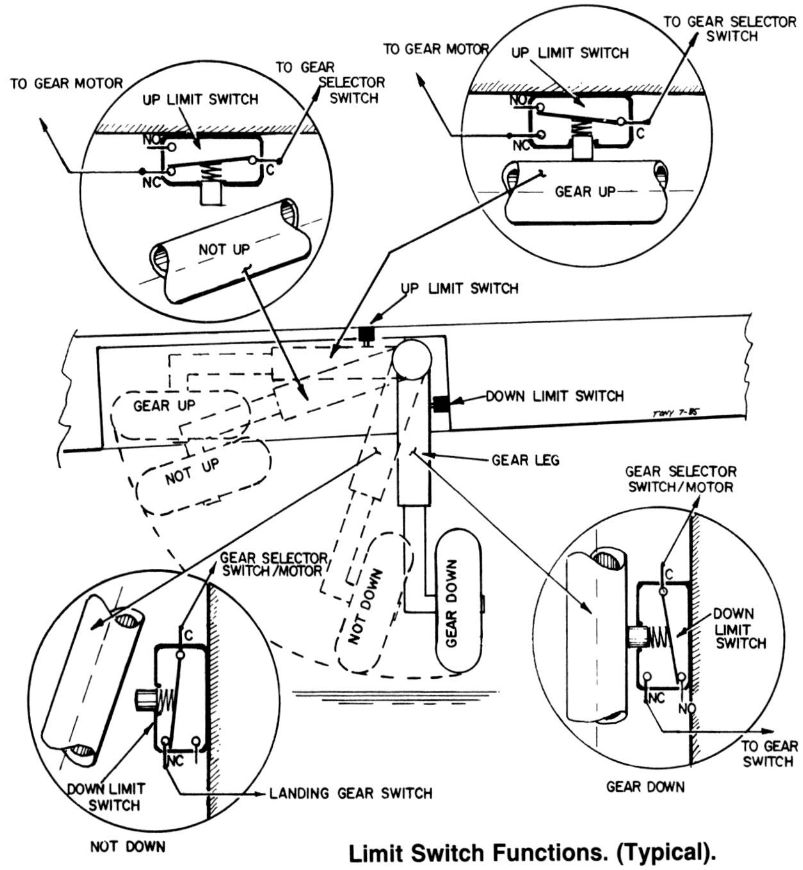

Limit Switch Functions. (Typical).

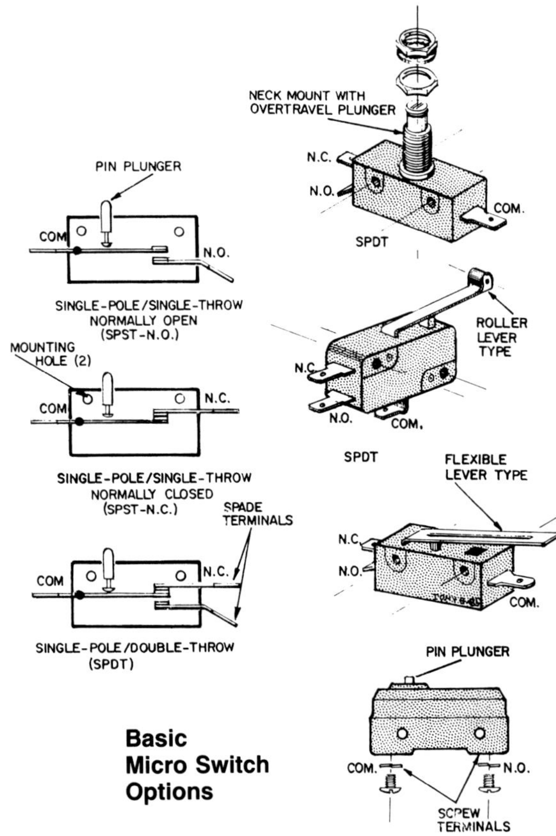

Basic Micro Switch Options

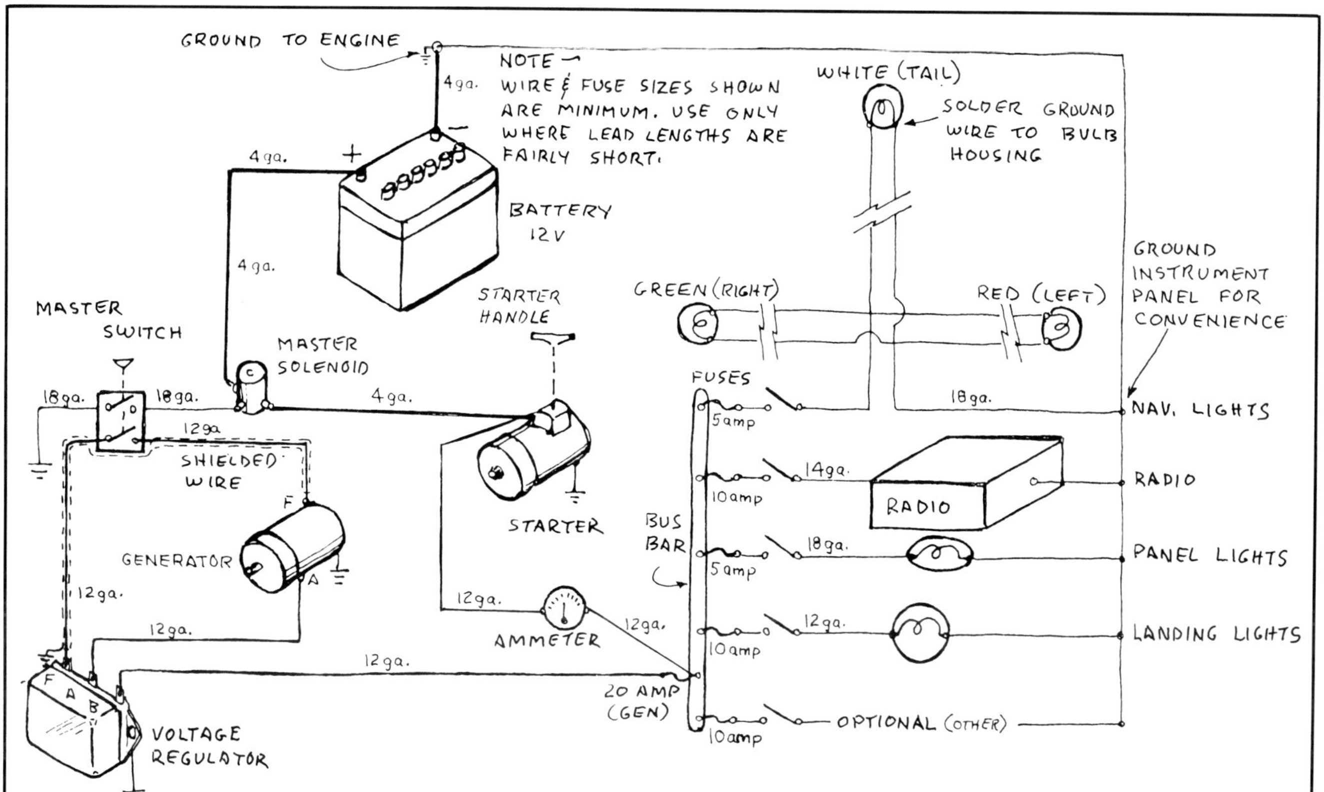

Basic Electrical System — Continental Engines (Two-Wire System for Wood Aircraft)

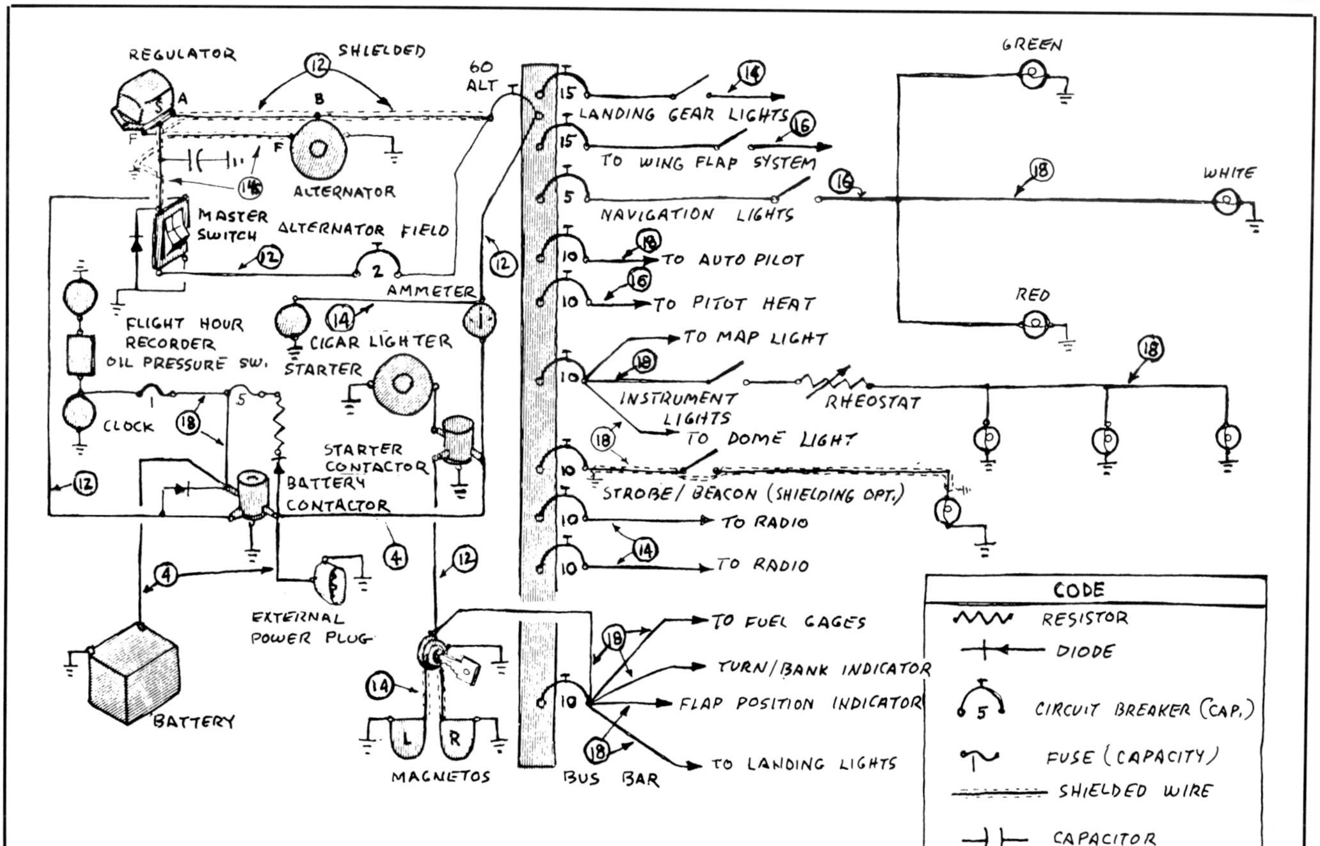

Power Distribution — Alternator System

Instrument Installation

22-1 INSTRUMENT INSTALLATION

The following recommendations are considered applicable to the installation of instruments when such instruments are required for particular operations.

22-2 AIRSPEED INDICATOR

This instrument should be so installed as to indicate true airspeed at sea level with maximum practicable accuracy.

22-3 MAGNETIC DIRECTION INDICATOR

Each magnetic direction indicator must be installed so that its accuracy is not excessively affected by the airplane's vibration or magnetic field.

The compensated installation should not have a deviation, in level flight, greater than ten degrees on any heading.

A placard must be installed on or near, the magnetic direction indicator.

The placard must show the calibration of the instrument in level flight with the engine(s) operating.

22-4 STATIC PRESSURE SYSTEM

Each instrument provided with static pressure case connections should be so vented that the influence of airplane speed, the opening and closing of windows, airflow variations, moisture, or other foreign matter will least affect the accuracy of the instruments.

22-5 GYROSCOPIC INSTRUMENTS

All gyroscopic instruments intended for operation under instrument flight rules should derive their energy from a power source of sufficient capacity to maintain their required accuracy in accordance with the manufacturers' recommendations. All instruments such as these designed to allow the amateur-built aircraft to fly at night or on instruments should be "yellow tagged," TSO'd or meet the requirements for the TSO. If instrument flying is contemplated, this must be indicated on the operating limitations from your FAA inspector.

22-6 INSTRUMENTS AND INSTALLATION

The minimum instrumentation should be as required by FAR Par. 91.33.

22-7 FLIGHT INSTRUMENTS (Required for VFR Flight)

Airspeed indicator.
Altimeter.
Magnetic direction indicator.

22-8 POWER PLANT INSTRUMENTS (Required for VFR Flight) - For Each Engine or Tank

Tachometer.
Fuel quantity indicator.
Fuel pressure gauge (for fuel pump-type installations).
Oil temperature indicator (not for two-cycle engines).
Oil pressure indicator (not for two-cycle engines).
Manifold pressure (variable pitch props).
Cylinder head temp. (aircraft with movable cowl flaps).

22-9 ARRANGEMENT AND VISIBILITY

Each flight, navigation and powerplant instrument for use by the pilot must be easily visible to him.

For each multiengine airplane, identical powerplant instruments must be located so as to prevent confusion as to which engine each instrument relates.

Instrument panel vibration should not be so severe as to damage, or impair the accuracy of any instrument.

VFR Day (FAR 91.33b) requires only: Airspeed, Altimeter, RPM, Oil Pressure and Temperature and Fuel gauge. Note: instrument markings, red line, green arc, yellow arc, red line.

Nicely laid out instrument panel. Normally open cockpit aircraft do not have such instrumentation — gyro compass, artificial horizon, turn coordinator, VOR/LOC indicator. Nice! Note radios mounted forward of stick. Side panels are great for electrical instruments, switches, fuses and phone jacks.

Above — A small section of the homebuilt parking area at the 1990 EAA Oshkosh convention. Most noticeable in this group are the tri-gear Glasairs. The Glasair offers impressive cruise speed with great cross country range and is very popular with homebuilders. Kits are available for conventional, tri-gear and retractable landing gear models.

Right — With the fuselage built of 4130 chrommoly steel tubing the Kolb Laser weighs only 350 lbs. empty. A unique design feature allows wing to fold "carrier style" for easy storage. A 4-cylinder 80 hp Rotax 912 4-stroke engine is installed. VW conversions can also be installed.

Below — The Wanderer, an original design by Larry Lam is powered with an 1800 cc VW engine. Aircraft is built around a wood sub-structure and covered with sculpted foam, epoxy and glass. Span is 20 ft., tapered in planform and thickness using a GAW-1 airfoil at the root and a GAW-2 at the tip. At 7000 ft. aircraft cruises at 185 mph.

This RotorWay Executive helicopter was built by Joe Colombo, Wayne, NJ. It won Grand Champion honors at the 1990 EAA Oshkosh convention — Colombo's second such honor, having taken a similar award with a RotorWay Scorpion helicopter several years ago. A rare accomplishment to win TWO Grand Champion awards!

Rotary Wings

A. C. Ballauer

Rotary wing aircraft — helicopters and gyroplanes fly according to the same aerodynamic principles as fixed wing aircraft, but because the rotary motion of the rotors is superimposed on forward motion of the entire craft, the aerodynamics and mechanics are much more complicated and involved.

This does not mean that these interesting and useful aircraft are beyond the amateur builder's means and abilities. On the contrary, if some simple fundamental rules are followed, successful and economical machines can be built.

A number of reference books are available for study of the principles of rotary wing flight and a review of these will show the prospective builder where problems are to be expected.

A study of the general proportions and other data of contemporary rotorcraft will guide the experimenter in the design of his craft and if such parameters as disc loading, rotor solidity, tip speed and power loading are kept in the range of these existing machines, there is no question as to whether the aircraft will fly or not. The real problems will be connected with handling characteristics and structural strength of the drive system and rotor systems.

Handling characteristics are difficult to discuss in a manual of this type because of the tremendous number of mechanical variations which are possible. Each variation has its effect on handling characteristics and stability and can be evaluated theoretically or experimentally on the completed machine. It is especially important to realize that no one has yet built a helicopter which is inherently stable in hovering and forward flight without the use of some automatic stabilizing equipment, independent of the rotor mechanism. In addition, it must be remembered that rotor control forces are a function of rate of control application and not stick position unless artificial "feel" is built in. Realization of the above factors and a cautious approach are the best advice in this area.

Determination of the structural integrity of the rotating parts is an extremely important problem, particularly since the major stresses are of an alternating character and fatigue failure of blades, etc., which would be a catastrophe, must be prevented. The drive system, hub parts, control system and rotor blades are all subject to steady and vibratory loads which vary greatly with speeds, maneuvers, power setting, etc. The state of the art of helicopter design has progressed to a point where it is possible to calculate these various loads, but the average experimenter will probably not be equipped, nor have the desire, to make these laborious calculations. In any case, good practice calls for experimental determination or verification of the stresses in the rotating mechanisms. This is usually done by strain gauge measurements of stresses in various parts while the aircraft is flown through a prescribed series of maneuvers. A comparison of the measured stresses and the previously determined fatigue life of the various parts of the aircraft permits an estimate of allowable life of the components to be estimated. In view of the need for fairly expensive and complicated equipment to accomplish a "flight vibratory strain survey," it is doubtful that many experimenters will carry through such a program. Since 50, or 100, or even several hundred hours of test flight time cannot substantiate a rotary wing design for any life beyond 75 percent of the test period, the problem of modifying the original experimental category limitations to permit less restricted use of the amateur-built aircraft, is a sticky one. This factor alone should receive thorough consideration by the experimenter before initiating his project.

All other design and construction problems closely parallel those of fixed wing aircraft and standard practices can be followed.

Federal Air Regulation Part 101 provides for training in "gyrogliders" attached to a vehicle on the surface of the earth.

FAR 101.1(d)(4)(b) states, "For the purpose of this part,

a gyroglider attached to a vehicle on the surface of the earth is considered to be a "kite".

As long as the gyroglider is operated below 150 feet above the surface of the earth it does not need to be registered. It can be used for solo instruction as a kite as long as it is in tow by a ground vehicle only. It is recommended that a CFI be on hand to observe all such flight and to log them for the solo flyer. If the pilot flies from 150 to 500 feet above the earth, he must give the closest FAA ATC (Air Traffic Control) facility that is nearest to the place of intended operation at least 24 hours before beginning the operation (a) the name and addresses of the owner(s) and operator(s), (b) the size and weight of the gyroglider, (c) the location of the operation, (d) the height above the surface of the earth at which the gyroglider is to be operated and, (e) the date, time and duration of the operation.

A waiver or authorization issued by the Administrator of the FAA is required for all operations which involve the following conditions: (a) towing in altitudes of 500 feet or more above the surface, or (b) within 5 miles of the boundary of the nearest airport, or (c) closer than 500 feet to the base of any cloud, or (d) during the hours of darkness unless gliding and marking requirements of 101.17 are observed, or (e) when the ground visibility is less than 3 miles.

The following is recommended for towed rotorcraft gliders:

1. A system of signals for communications between the operator and the driver of the tow vehicle is considered necessary.
2. A reasonably smooth and level surface at least 1,000 feet long is desirable to provide room for take-off, straight flight, and landing.
3. The operator should make the ground practice tests recommended in the kit or construction plan instructions before attempting flight operations.
4. The glider should be equipped with a means for quickly detaching the tow cable in case of emergency.
5. If flights are made at airports, the operator should coordinate all operations with the appropriate airport officials and be familiar with the airport rules and procedures.
6. Towing by aircraft is not permitted under FAR 101, and should not be attempted.
7. The operator should not attempt free flight until he is thoroughly familiar with the handling of the aircraft and has made certain that the appropriate aircraft and pilot certificates are held.

For gyroplanes, taxi tests should be conducted to determine the handling characteristics of the aircraft prior to attempting flight operations, initial test flights should be made from airports with runways of sufficient length to determine take-off and landing characteristics of the aircraft without undue risks to the operator, and that flight tests away from airports must be made within the boundaries of approved flight test areas.

For helicopters, we offer this advice. Stability, vibration, and balance tests should be completed prior to attempting hover or flight operations. Also that hover tests should be completed prior to attempting forward flight.

Finally, a list of general safety hints for all rotorcraft operations is offered.

1. Airplane pilots should be prepared to cope with a non-conventional aircraft which will not possess the stable characteristics of a conventional airplane.
2. The effect of collective pitch or cyclic control movements may cause confusion to the airplane pilot and should be thoroughly understood by all operators.
3. Operators of rotorcraft having three or more rotor blades should be particularly cautious of ground resonance, a condition of rotor unbalance that if allowed to continue, will usually result in structural failure.

Single place Helicraft Skylark I helicopter is powered with a Rotax 582 64 hp engine. With a rotor diameter of 19 ft., it has a disc loading of 2.5 lbs. per sq. in. Empty weight is 325 lbs., gross 700 lbs. Cruise speed 70 mph.

Gyrocopters have been popular since the earliest days of EAA when Bensen gyrocopters were built in greater numbers than any other aircraft. A popular airshow performer at EAA's convention for many years is Ken Brock, shown above preflighting his KB-2 prior to takeoff. The KB-2 is assembled from eight self-contained kits and can be powered with a McCulloch or VW powerplant.

The regulations pertaining to rotorcraft and all aircraft are contained in the Code of Federal Regulations, Part 14. These regulations are broken down into four parts:

Aeronautics & Space Part 1 - 59
Aeronautics & Space Part 60 - 139
Aeronautics & Space Part 140 - 199
Aeronautics & Space Part 200 - 1199

These regulations are the equivalent of the FAR's with the exception that they are updated only once per year. They are for sale by:

Superintendent of Documents
U.S. Government Printing Office
Washington, DC 20402

or your local G.P.O. Bookstore.

FAR Part 61 gives general information on pilot's licenses and specific information on rotorcraft. FAR 61.87(c)(2), 61.87(c)(3), 61.87(d) and 61.87(d)(1). In this particular instance, 61.87(d)(1) states "has given him instruction **in** the make and model of aircraft in which the solo flight is made". Please note this does not say "flight instruction". Further specific information on rotorcraft is covered in FAR 61.105(b) and 61.107(c). 61.113 covers rotorcraft ratings and aeronautical experience.

It is possible for a flight instructor under this part to observe a gyrocopter in solo flight from the ground for endorsing his log book as OK for solo flight. In addition, FAR Part 101, specifically 101.1(4)(b) states "For the purpose of this part, a gyroglider attached to a vehicle on the surface of the earth is considered to be a kite." FAR Part 101.13 operating limitations states (a) except as provided in paragraph (b) of this section, no person may operate a moored balloon or kite - (1) less than 500 feet from the base of any cloud, (2) more than 500 feet above the surface of the earth, (3) from an area where the ground visibility is less than 3 miles; or (4) within 5 miles of the boundary of any airport. (b) Paragraph (a) of this section does not apply to the operation of a balloon or kite below the top of any structure and within 250 feet of it, if that shielded operation does not obscure any lighting on the structure. FAR 101.15 "Notice Requirements": No person may operate an unshielded moored balloon or kite more than 150 feet above the surface of the earth unless, at least 24 hours before beginning the operation, he gives the following information to the FAA ATC facility that is nearest to the place of intended operation: (a) The names and addresses of the owners and operators. (b) The size of the **balloon** or the size and weight of the kite. (c) The location of the operation. (d) The height above the surface of the earth at which the balloon or kite is to be operated. (e) The date, time and duration of the operation. Further information in the FAR 101 covers additional requirements.

For a private pilot's (gyrocopter) license, a gyrocopter pilot has to have a total of 40 hours dual and solo, including 20 hours of flight instruction with at least 10 hours of solo flight time in gyroplanes, as covered in 61.113.

Part 61 also covers the requirements for private pilot (helicopter) ratings: For a helicopter rating, an applicant must have at least a total of 40 hours of flight instruction and solo flight time in aircraft with at least 15 hours of that solo flight time in helicopters, which must include specific requirements covered in 61.113.

It is hoped that the above explanations will answer many of your questions that rotorcraft owners have. Purchase of the FARs and study is recommended to the potential rotorcraft user.

Part 103 - Ultralight Vehicles

On October 4, 1982, part 103 of the Federal Aviation Regulations (FAR) became effective to govern the operations of ultralight vehicles in the United States. The rules governing ultralight vehicles were needed to achieve an acceptable level of safety by reducing potential conflict with other airspace users and by protecting persons and property on the ground. The intent was to provide an acceptable level of safety with a minimum amount of regulation.

The rule defined ultralights and restricted their operations in certain categories of airspace as well as limiting other aspects of their operations. It was believed at the time that an acceptable level of safety could be achieved without aircraft certification and registration requirements. Similarly, part 103 exempted pilots of ultralight vehicles from the requirement to hold a pilot or medical certificate.

Subpart A — General

Sec.
103.1 Applicability.
103.3 Inspection requirements.
103.5 Waivers.
103.7 Certification and registration.

Subpart B — Operating Rules

103.9 Hazardous operations.
103.11 Daylight operations.
103.13 Operation near aircraft: right-of-way rules.
103.15 Operations over congested areas.
103.17 Operations in certain airspace.
103.19 Operations in prohibited or restricted areas.
103.21 Visual reference to the surface.
103.23 Flight visibility and cloud clearance requirements.

Authority: Secs. 307.313(a), 601(a), 602, and 603, Federal Aviation Act of 1958 (49 U.S.C. 1348, 1354(a), 1421(a), 1422, and 1423); sec. 6(c), Department of Transportation Act (49 U.S.C. 1655(c)

SUBPART A — GENERAL

103.1 Applicability.

This part prescribes rules governing the operation of ultralight vehicles in the United States. For the purposes of this part, an ultralight vehicle is a vehicle that:

(a) Is used or intended to be used for manned operation in the air by a single occupant;

(b) Is used or intended to be used for recreation or sport purposes only;

(c) Does not have any U.S. or foreign airworthiness certificate; and

(d) If unpowered, weighs less than 155 pounds: or

(e) If powered:

(1) Weighs less than 254 pounds empty weight, excluding floats and safety devices which are intended for deployment in a potentially catastrophic situation;

(2) Has fuel capacity not exceeding 5 U.S. gallons;

(3) Is not capable of more than 55 knots calibrated airspeed at full power in level flight (63.5 m.p.h.); and

(4) Has a power-off stall speed which does not exceed 24 knots calibrated airspeed. (27.7 m.p.h.).

103.3 Inspection requirements.

(a) Any person operating an ultralight vehicle under this part shall, upon request, allow the Administrator, or his designee, to inspect the vehicle to determine the applicability of this part.

(b) The pilot or operator of an ultralight vehicle must, upon request of the Administrator, furnish satisfactory evidence that the vehicle is subject only to the provisions of this part.

103.5 Waivers.

No person may conduct operations that require a deviation from this part except under a written waiver issued by the Administrator.

103.7 Certification and registration.

(a) Notwithstanding any other section pertaining to certification of aircraft or their parts or equipment, ultralight vehicles and their component parts and equipment are not required to meet the airworthiness certification standards specified for aircraft or to have certificates of airworthiness.

(b) Notwithstanding any other section pertaining to airman certification, operators of ultralight vehicles are not required to meet any aeronautical knowledge, age, or experience requirements to operate those vehicles or to have airman or medical certificates.

(c) Notwithstanding any other section pertaining to registration and marking of aircraft, ultralight vehicles are not required to be registered or to bear markings of any type.

SUBPART B — OPERATING RULES

103.9 Hazardous operations.

(a) No person may operate any ultralight vehicle in a manner that creates a hazard to other persons or property.

(b) No person may allow an object to be dropped from an ultralight vehicle if such action creates a hazard to other persons or property.

103.11 Daylight operations.

(a) No person may operate an ultralight vehicle except between the hours of sunrise and sunset.

(b) Notwithstanding paragraph (a) of this section, ultralight vehicles may be operated during the twilight periods 30 minutes before official sunrise and 30 minutes after official sunset or, in Alaska, during the period of civil twilight as defined in the Air Almanac, if:

(1) The vehicle is equipped with an operating anticollision light visible for at least 3 statute miles; and

(2) All operations are conducted in uncontrolled airspace.

103.13 Operation near aircraft; Right-of-way rules.

(a) Each person operating an ultralight vehicle shall maintain vigilance so as to see and avoid aircraft and shall yield the right-of-way to all aircraft.

(b) No person may operate an ultralight vehicle in a manner that creates a collision hazard with respect to any aircraft.

(c) Powered ultralights shall yield the right-of-way to unpowered ultralights.

103.15 Operations over congested areas.

No person may operate an ultralight vehicle over any congested area of a city, town, or settlement, or over any open air assembly of persons.

103.17 Operations in certain airspace.

No person may operate an ultralight vehicle within an airport traffic area, control zone, terminal control area, or positive control area unless that person has prior authorization from the air traffic control facility having jurisdiction over that airspace.

103.19 Operations in prohibited or restricted areas.

No person may operate an ultralight vehicle in prohibited or restricted areas unless that person has permission from the using or controlling agency, as appropriate.

103.21 Visual reference with the surface.

No person may operate an ultralight vehicle except by visual reference with the surface.

103.23 Flight visibility and cloud clearance requirements.

No person may operate an ultralight vehicle when the flight visibility or distance from clouds is less than that in the following table, as appropriate:

Flight altitudes	Minimum flight visibility[1]	Minimum distance from clouds
1,200 feet or less above the surface regardless of MSL altitude.		
(1) Within controlled airspace	3	500 feet below, 1,000 feet above, 2,000 feet horizontal.
(2) Outside controlled airspace	1	Clear of clouds.
More than 1,200 feet above the surface but less than 10,000 feet MSL.		
(1) Within controlled airspace	3	500 feet below, 1,000 feet above, 2,000 feet horizontal.
(2) Outside controlled airspace	1	500 feet below, 1,000 feet above, 2,000 feet horizontal.
More than 1,200 feet above the surface and at or above 10,000 feet MSL.	5	1,000 feet below, 1,000 feet above, 1 statute mile horizontal.

[1]Statute miles.

Miscellaneous Recommendations

24-1 LEVELING MEANS

A suitable location on the aircraft should be established for the purpose of determining when the aircraft is in a level position. In most cases any arbitrary reference may be used such as a longeron, the cockpit floor or any line between two points suitable for that particular design.

24-2 JACKING MEANS

Considerations should be given to the need for building into the aircraft or providing by some other means, safe jacking points for the aircraft.

24-3 STANDARD WEIGHTS

In computing weights for any purpose the following standard values should be used:

Pilot and Passenger	170 lbs/person
Parachute	20 lbs each
Gasoline	6 lbs/gallon
Oil	7.5 lbs/gallon
Water	8.5 lbs/gallon
Aerobatic Pilot with Parachute	190 lbs/person

24-4 TIE DOWN POINTS

Suitable points should be provided whereby the aircraft can be safely moored or tied down. It is customary to provide three tie down points. The wing tie down points are best located at least ½ of the span outboard.

24-5 SHOULDER HARNESS PLEDGE

It is earnestly recommended that all builders comply with the spirit and intent of the EAA Shoulder Harness Pledge stated as follows:

"I *(your name)* hereby promise to install and wear shoulder harness and safety belts in my amateur-built aircraft to protect myself, passenger and the good name of the Experimental Aircraft Association. Air Force and Navy tests have proven that a 20G harness will eliminate 90% of aircraft accident injuries."

24-6 EMERGENCY LOCATER TRANSMITTER

Reference is AC 91-44A.

Experience has shown that ELTs should be attached preferably to airframes or other solid structures.

ELT antennas should be located as far as practical from other installed antennas. Although most recommendations are to locate the ELT as far aft as practical on the airframe, this may not be too good an idea for some homebuilts that may have an aft CG problem already.

ELTs should not be attached to unreinforced panels such as the sides of baggage compartments as the vibration of the panel may activate the ELT.

24-7 RADIO STATION LICENSE

The Federal Communications Commission regulates the use of all radio transmitters. If you intend to install and equip your airplane with one you will be required to obtain a Radio Station License from the FCC. To accomplish this requirement you must obtain and complete FCC Form 404-2, enclose the required fee, if required, and mail it to the FCC.

There are no particular technical qualifications required of the builder in obtaining this license. Forms may be obtained through FCC offices only . . . not from the FAA or the EAA:

FEDERAL COMMUNICATIONS COMMISSION
Gettysburg, Pennsylvania 17325

NOTE: ELT's without voice capability do not require an FCC station license. ELT's with voice capability do require an FAA license (Voice transmitter). ELT's are not required for training flights within 50 miles of the home base, and are not required for single-place aircraft nor during test periods.

NOTE: Hand held radios **do** require an aircraft station license. The FCC license for the pilot is not required unless he goes outside the continental limits of the United States.

Above — Single place, all-wood, Corby Starlet, an Austrailian design, is fitted with an 1835cc VW conversion engine. This outstanding Starlet was built by Ray Downs, Costa Mesa, CA. Aircraft weighs 509 lbs. and cruises at 135 mph. Downs previously had built a trophy winning Wittman Tailwind.

Left — KR-2 is a durable design that dates back to the early '70s. Over 10,000 sets of plans for this and the single place KR-1 have been sold. Basic structure of wood with foam/Dynel/epoxy secondary structure permits easy shaping of surfaces and builders have modified these aircraft extensively. Originally designed with a retractable gear, this KR-2 built by Louie Brochetti, Carlsbad, CA sports a fixed gear.

Below — a 2/3 scale P-51 Mustang by Charlie Armstrong, Midland, TX. Aircraft has a welded steel tube fuselage with fiberglass outer shell, and all metal wings and tail. Power is a Fred Geschwender conversion Ford engine of 358 hp. Aircraft has a top speed of 240 mph at sea level.

The 51% Rule

(Major Portion Rule)

Since the very early 1950s and the establishment of the amateur built aircraft program as a part of the experimental category, the well known 51% rule has been in effect. This means that a homebuilt aircraft constructed by an individual or a group of individuals for educational or recreational purposes would qualify for the amateur built category with certain operating limitations, which over the years has been lessened due to earned safety considerations.

The question has often been asked by the commercial interests of the movement . . . "why 51% for those wishing to market kits or completed aircraft under this program"?

According to the retired FAA men and long-time EAA members who were instrumental in the establishment of the CAA, they wanted to insure that the builder constructed over one-half of the aircraft. Fifty-one percent is obviously over one-half. This then also satisifed those in CAA who were not entirely in favor of such a program.

Since that time because of EAA's involvement in the movement greater freedoms have been earned by the movement.

Presented here in its entirety is another refinement of the rule that gives a bit more latitude to the movement while working with the EAA.

We would like to thank recently retired FAA'er George Pour, Manager of the Aircraft Manufacturing Division, and his associates for their efforts in preparing the following guidelines.

Department of Transportation
Federal Aviation Administration
Order No. 8130.18

SUBJ: MAJOR PORTION EVALUATION CHECKLISTS FOR AMATEUR-BUILT AIRCRAFT

1. **Purpose.** This order provides checklists for determining whether an aircraft is eligible for airworthiness certification under Federal Aviation Regulations (FAR) 21.191(g).

2. **Distribution.** This order is distributed to the branch level in the Offices of Flight Operations and Airworthiness; to the section level in the Aircraft Certification Divisions; to the branch level in the Flight Standards Divisions and the Aviation Standards National Field Office to the Airworthiness Section at the FAA Academy; to all Aircraft Certification Offices; Aircraft Certification Field Offices; Manufacturing Inspection District/Satellite Offices; and to all General Aviation and Flight Standards District Offices.

3. **Background.**

a. Under FAR 21.191(g) an aircraft is eligible for certification as amateur built when the "major portion" has been fabricated and assembled by persons who undertook the construction project solely for their own education or recreation. Over the years, since the rule was promulgated, questions and controversy have frequently arisen with respect to defining what constitutes the major portion of an aircraft, particularly in the case of aircraft built from kits. Advisory Circular (AC 20-27) Certification and Operation of Amateur-Built Aircraft, and subsequent revisions, established "more than 50 percent" as the major portion, but more definitive criteria were not provided. It was left to the judgment of Federal Aviation Administration (FAA) inspectors who evaluate an amateur-built aircraft project to establish whether the builder fabricated the major portion. Such personal judgments varied considerably among inspectors, which resulted in a lack of nationwide uniformity in administering FAR 21.191(g). The need for a better means of evaluating amateur-built aircraft construction projects became increasingly acute with the advent of small aircraft built from composite materials, where many major components must be fabricated in high-cost molds by the kit manufacturers.

b. Notice N8130.39 was issued July 25, 1983, to establish a trial program for the evaluation of a checklist concept for use in determining whether an aircraft is eligible for airworthiness certification under FAR 21.191(g). The notice provided checklists in Appendices 1 and 2 in an effort to establish uniformity in implementing FAR 21.191(g), by providing fixed criteria for determining whether a builder fabricated and assembled the major portion of an aircraft for education or recreation, or for determining whether an aircraft built from a kit is eligible for certification under FAR 21.191(g). The premise upon which the checklists were based was that, for a builder to derive education or recreation from an aircraft building project, the builder should learn to use the tools and perform the functions that are generally needed to build a small aircraft. The checklists contained most of the basic tools and functions that may be required for a complete aircraft. However, in keeping with the major portion criteria of FAR 21.191(g), more than half of the items included in the checklists would need to be used or performed by the builder of the aircraft in order for the aircraft to be eligible for certification as amateur built. Notice N8130.39 expired on July 1, 1984. No adverse comments were received from any source, even though feedback from FAA offices indicated that the checklists were used extensively in evaluating kits, and in a few cases for settling questions concerning "major portion". This order, therefore, is issued to permanently establish the checklist concept evaluated through Notice N8130.39.

4. **Use of Checklists.**

a. FAA Field Offices involved in amateur-built aircraft certification may use the checklists in Appendices 1 and 2 in the following situations:

(1) To determine whether a kit from which an aircraft is to be built would meet the major portion provisions of FAR 21.191(g).

(2) To settle any controversy with respect to major portion that may arise in the certification of an amateur-built aircraft under the provisions of AC 20-27. The use of a checklist should normally not be necessary for an aircraft built from a kit previously found eligible for amateur-built certification, or where the builder's records, data, and notarized statement provide ample proof that the builder fabricated the major portion. Questionable cases might be where the aircraft was built from prefabricated major components that are readily available from aircraft parts suppliers, or the aircraft was built from a kit that had not been found eligible by the FAA.

(3) To provide guidance to a prospective builder as to what constitutes the major portion before beginning the project.

(4) To provide guidance to a kit manufacturer desiring to determine if a proposed kit would meet the major portion provision of FAR 21.191(g). By completion of the applicable checklist, it may be determined without a prototype if a proposed kit would be eligible for amateur-built certification. If not, the parts and materials in the kit could be "tailored" to comply by increasing the number of tools and/or functions that the builder would need to use to complete the aircraft.

(5) To provide guidance to Designated Airworthiness Representative or Experimental Aircraft Association designees who are involved in certification of amateur-built aircraft.

Primed and with most assemblies installed. When turtledeck is bolted to the structure covering will begin. It's great to have enough work space to fully assemble the aircraft so that rigging and alignment can be checked before covering is started.

b. Appendices 1 and 2 may be locally reproduced for distribution as needed.

Appendix 1. Major Portion Evaluation Checklist Amateur-Built Aircraft Over 350 Lbs. Maximum Empty Weight

The purpose of this checklist is to establish whether the major portion of an amateur-built aircraft was fabricated and assembled by the builder. When repetitive functions are required to complete a system or assembly, such as ribs for the wings, the builder may use a helper(s); or, in the case of a "kit" aircraft, less than half of the items may be prefabricated and provided with the kit. For example, if the wings for a given aircraft use 24 ribs, the kit may contain 11 prefabricated ribs and materials for the other 13; or, if 30 fairlead/pulley brackets of similar construction are required for the flight control system, the kit may contain 14 prefabricated brackets, and materials for the other 16.

The number derived upon completion of the checklist indicates the percentage of the aircraft fabricated and assembled by the builder for education or recreation. For the purpose of compliance with FAR 21.191(g), the number should be more than 50.

NOTE: The term "fabricate" means that the builder performs the functions needed to make a completed part(s) from raw materials.

Instructions

1. Circle or mark the number preceding each tool and function that is used by the builder to complete the aircraft.
2. Count all items marked.

Total ___ .

3. Divide the total marked by 0.85.

Resultant percentage ___ .

Tools Total marked ___ .

1. Arc welder.
2. Bandsaw/jigsaw.
3. Blueprints, drawings, manuals.
4. Cable tensiometer.
5. Cable swaging/nico press.
6. Chisels, metal working.
7. Chisels, wood working.
8. Drillpress.
9. Drill motors (hand).
10. Files (wood/metal working).
11. Gas welder.
12. Grinders (electric).
13. Hacksaw.
14. Lathe (wood).
15. Lathe (metal).
16. Paint spray gun.
17. Planer/shaper (metal).
18. Planer (wood).
19. Plastic foam hot wire cutters.
20. Pop rivet gun.
21. Rivet gun/bucking bars.
22. Ribstitching needles.
23. Rigging protractor/templates.
24. Sewing machine.
25. Sheetmetal shears (mechanical).
26. Sheetmetal shears (hand).
27. Sheetmetal bending brake.
28. Sheetmetal forming tools.
29. Timing equipment, engine/magneto.
30. Torque wrench.
31. Tube cutter/bender.
32. Tube flaring tools.
33. Wire lugger.
34. Woodsaw (hand).
35. Woodsaw (electrical).
36. Woodworking clamps.

Empennage/Canard Total marked ___ .

1. Cut and fit steel tube joints/clusters.
2. Weld assemblies.
3. Cut and fit wood members.
4. Glue wood members.
5. Cut/shape foam material.
6. Fabricate brackets/fittings/hinges.
7. Install covering (fabric, metal, plywood).
8. Paint finished components.
9. Install empennage/canard on fuselage.
10. Rig control surfaces.
11. Form sheet metal parts.

Complete Aircraft Total marked ___ .

1. Weigh for weight and balance calculation.
2. Develop weight and balance report.

Appendix 2. Major Portion Evaluation Checklist Amateur-Built Aircraft Up To 350 Lbs. Maximum Empty Weight

The purpose of this checklist is to establish whether the major portion of an amateur-built aircraft was fabricated and assembled by the builder. When repetitive functions are required to complete a system or assembly, such as ribs for the wings, the builder may use a helper(s); or, in the case of a "kit" aircraft, less than half of the items may be prefabricated and provided with the kit. For example, if the wings for a given aircraft use 24 ribs, the kit may contain 11 prefabricated ribs and materials for the other 13; or, if 30 fairlead/pulley brackets of similar construction are required for the flight control system, the kit may contain 14 prefabricated brackets, and materials for the other 16.

The number derived upon completion of the checklist indicates the percentage of the aircraft fabricated and assembled by the builder for education or recreation. For the purpose of compliance with FAR 21.191(g), the number should be more than 50.

NOTE: The term "fabricate" means that the builder performs the functions needed to make a completed part(s) from raw materials.

Instructions

1. Circle or mark the number preceding each tool and function that is used by the builder to complete the aircraft.
2. Count all items marked.

Total ___ .

3. Divide the total marked by 0.65.

Resultant percentage ___ .

Tools Total marked ___ .

1. Arc welder.
2. Bandsaw/jigsaw.
3. Blueprints, drawings, manuals.
4. Cable swaging/nico press.
5. Cable cutter.
6. Drillpress.
7. Drill motors (hand).
8. Deburring tools.
9. Files.
10. Gas welder.
11. Grinders (electric).
12. Hacksaw.
13. Paint spray gun.
14. Riveting tools.
15. Rigging tools.
16. Sewing machine.
17. Sheet metal tools.
18. Soldering iron.
19. Timing equipment.
20. Torque wrench.
21. Tools, special purpose.
22. Volts/ohmmeter.
23. Woodsaw.
24. Tube bender.

Functions Total marked ___ .

Fuselage Total marked ___ .

1. Fabricate assembly jig(s).
2. Fabricate special tools.
3. Cut and fit fuselage members.
4. Fabricate landing gear components
5. Fabricate & install brackets/fittings for all systems.
6. Measure/cut control cables/tubes.

Many builders use a two part epoxy primer to preserve tubing and metal surfaces against rust and corrosion.

7. Swage cable fittings.
8. Assemble fuselage structure.
9. Fabricate & install cockpit flight control components.
10. Fabricate & install powerplant controls.
11. Fabricate & install cockpit/cabin interior finish, incl. seats.
12. Form sheet metal parts.
13. Install fuselage covering (fabric, metal, plywood).
14. Paint finished fuselage.

Powerplant Total marked ___ .

1. Cut and fit steel tube joints/clusters.
2. Weld engine mount.
3. Install engine mount and engine.
4. Fabricate & install brackets & fittings.
5. Install engine controls.
6. Fabricate & install fuel tank.
7. Install fuel system.
8. Install and track propeller.
9. Adjust/tune engine.

Wings (Sails) Total marked ___ .

1. Fabricate wing supoort jig.
2. Fabricate wing spars.
3. Fabricate wing ribs.
4. Fabricate brackets and fittings.
5. Fabricate & install truss members/wires.
6. Assemble wing structure.
7. Install wing covering.
8. Install rib stitching.
9. Measure/cut control cables/tubes.
10. Fabricate ailerons (spoilerons) and attach fittings.
11. Fabricate wing attach fittings.
12. Install and rig wings to fuselage.
13. Rig aileron (spoileron) controls.

Empennage/Canard Total marked ___ .

1. Fabricate jigs/special tools.
2. Cut and fit empennage members.
3. Assemble empennage members.
4. Fabricate brackets, fittings, hinges.
5. Cover empennage structure.
6. Install empennage/canard to fuselage.
7. Rig control surfaces.

Complete Aircraft Total marked ___ .

1. Weigh for weight and balance calculation.
2. Develop weight and balance report.

Left — Steel tube Cuby fuselage that was built from a kit. Aircraft are designed for 65 to 100 hp Continentals. Note minimum wood fairing strips. The success of this replica Cuby, like the original Cub, is its' great simplicty!

Below — Composite construction employs new techniques in building. In the past decade this type construction has proven extremely popular with homebuilders. Compound curves no problem. A Task Research team member below works on the Silhouette fuselage that was constructed during an Oshkosh workshop.

Above — Many all wood aircraft are under construction and for anyone who has worked with wood, it's usually the choice when selecting an aircraft design. Some very high performance homebuilts are all wood.

Right — And some prefer to work with metal! Many designs of all metal aircraft are available to homebuilders. Note rows of Clecos holding metal bottom skin prior to riveting.

Wood, Aluminum, Steel And Composites

. . . and the Properties of each

by Chris Heintz

Aircraft structures are basically *unidirectional.* This means that one dimension, the length, is much larger than the others — width or height. For example, the span of the wing and tail spars is much longer than their width and depth; the ribs have a much larger chord length than height and/or width; a whole wing has a span that is larger than its chords or thickness; and the fuselage is much longer than it is wide or high. Even a propeller has a diameter much larger than its blade width and thickness, etc. . . . For this simple reason, a designer chooses to use unidirectional material when designing for an efficient strength to weight structure.

Unidirectional materials are basically composed of thin, relatively flexible, long fibers which are very strong in tension (like a thread, a rope, a stranded steel wire cable, etc.)

An aircraft structure is also very close to a *symmetrical structure.* That means the up and down loads are almost equal to each other. The tail loads may be down or up depending on the pilot raising or dipping the nose of the aircraft by pulling or pushing the pitch control; the rudder may be deflected to the right as well as to the left (side loads on the fuselage). The gusts hitting the wing may be positive or negative, giving the up or down loads which the occupant experiences by being pushed down in the seat . . . or hanging in the belt.

Because of these factors, the designer has to use a structural material that can withstand both tension and compression. Unidirectional fibers may be excellent in tension, but due to their small cross section, they have very little inertia (we will explain inertia another time) and cannot take much compression. They will escape the load by buckling away. As in the illustration, you cannot load a string, or wire, or chain in compression.

In order to make thin fibers strong in compression, they are "glued together" with some kind of an "embedding". In this way we can take advantage of their tension strength and are no longer penalized by their individual compression weakness because, as a whole, they become compression resistant as they help each other to not buckle away. The embedding is usually a lighter, softer "resin" holding the fibers together and enabling them to take the required compression loads. This is a very good structural material.

WOOD

Historically, *wood* has been used as the first unidirectional structural raw material. Nature, in her wisdom, has provided a beautiful unidirectional material by making certain trees grow in certain conditions: They have to be tall and straight and their wood must be strong and light. The cross section of a tree trunk shows the "annual rings" (a ring per year so that we can "count" the tree's age). The dark bands (late wood) contain many fibers, whereas the light bands (early wood) contain much more "resin". Thus the wider the dark bands, the stronger and heavier the wood. If the dark bands are very narrow and the light bands quite wide, the wood is light but not very strong. To get the most efficient strength to weight ratio for wood we need a definite numbers of bands per inch (see ANC No. 18, 1951). In fact, what we want is a good balance of "early" and "late" wood, or in other words, very special growing conditions, i.e., the geographic altitude where the trees grow varies with the latitude and local climatic conditions. Although this is a very interesting subject we will not go further into such details except to mention that it is nature who supplies us with a very efficient material from its plant kingdom. Remember that contrary to the strictly mineral world hopelessly subject to gravity pulling everything down, the plant has a force within itself which makes it grow against gravity, upwards. If we could use those life forces in our machines we could lift off without the help of an engine . . . Aviation still has a lot to discover . . .

Another subject we will not deal with here is the testing of wood. There are a few simple tests (humidity, dynamics, resilience) but it seems that nobody knows them anymore.

Some of our aircraft structures are two-dimensional (length and width are large with respect to thickness). Plywood is often used for such structures. Several thin boards (foils) are glued together so that the fibers of the various layers cross over at different angles (usually 90 degrees today — years back you could get them at 30 and 45 degrees as well). Plywood makes excellent "shear webs" if the designer knows how to use plywood efficiently. (We will learn the basis of stress analysis sometime later.)

To close this discussion on wood, let us plainly state the fact that our present day bureaucratic civilization uses so much paper that we are depleting the planet of trees without replanting them correctly. Today good aircraft wood is very hard to come by. Instead of using one good board for our spars, we have to use laminations because large pieces of wood are practically unavailable, and we no longer can trust the wood quality; we have to use may laminations so that the "average" has a reasonable chance to give us the required strength without too much penalty from a weight standpoint. From an availability point of view, we simply need a substitute for what nature has supplied us with until now.

ALUMINUM ALLOYS

So, since wood may not be as available as it was before, we look at another material which is *strong, light* and *easily available* at a *reasonable price* (there's no point in discussing Titanium — it's simply too expensive). *Aluminum alloys* are certainly one answer. We will discuss the properties of those alloys which are used in light plane construction in more detail later. For the time being we will look at aluminum as a construction material.

Extruded Aluminum Alloys

Due to the manufacturing process for aluminum we get a unidirectional material quite a bit stronger in the lengthwise direction than across. And even better, it is not only strong in tension but also in compression. Comparing extrusions to wood, the tension and compression characteristics are practically the same for aluminum alloys so that the linear stress analysis applies. Wood, on the other hand, has a tensile strength about twice as great as its compression strength;

accordingly, special stress analysis methods must be used and a good understanding of wood under stress is essential if stress concentrations are to be avoided!

Aluminum alloys, in thin sheets (.016 to .125 of an inch) provide an excellent two dimensional material used extensively as shear webs — with or without stiffeners — and also as tension/compression members when suitably formed (bent!).

It is worthwhile to remember that aluminum is an artificial metal. There is no aluminum ore in nature. Aluminum is manufactured by applying electric power to bauxite (aluminum oxyde) to obtain the metal, which is then mixed with various strength-giving additives. (In a later article, we will see which additives are used, and why and how we can increase aluminum's strength by cold work hardening or by tempering.) All the commonly used aluminum alloys are available from the shelf of dealers. When requested with the purchase, you can obtain a "mill test report" that guarantees the chemical and physical properties as tested to accepted specifications. (MIL standards, QQA250 — XYZ).

As a rule of thumb, aluminum is three times heavier, but also three times stronger than wood. Steel is again three times heavier and stronger than aluminum.

STEEL

The next material to be considered for aircraft structure will thus be steel, which has the same weight-to-strength ratio of wood or aluminum.

Apart from mild steel which is used for brackets needing little strength, we are mainly using a chrome-molybdenum alloy called AISI 4130N or 4140. (AISI .1025 is no longer available.)

The common raw materials available are tubes and sheet metal. Steel, due to its high density, is not used as shear webs like aluminum sheets or plywood. Where we would need say a .100" plywood, a .032 inch aluminum sheet would be required, but only a .010 steel sheet would be required, which is just too thin to handle with any hope of a nice finish. That is why a steel fuselage uses tubes also as diagonals to carry the shear in compression or tension and the whole structure is then covered with fabric (light weight) to give it the required aerodynamic shape or desired look. It must be noted that this method involves two techniques: steel work **and** fabric covering.

The advantage of 4130N steel structure is that it can readily be welded together. This applies especially in North America where the welder does not have to be "approved" as he has to be in Europe and Australia. This difference in regulations, historically, has to do with the "pioneer spirit" and explains why welded steel fuselages are so common here and practically nowhere else.

We will be discussing tubes and welded steel structures in more detail later and go now to "artificial wood" or composite structures.

COMPOSITE MATERIALS

The designer of composite aircraft simply uses fibers in the desired direction exactly where and in the amount required. The *fibers* are embedded in *resin* to hold them in place and provide the required support against buckling. Instead of plywood or sheet metal which allows single curvature only, the composite designer uses cloth where the fibers are laid in two directions (the woven thread and weft) also embedded in resin. This has the advantage of freedom of shape in double curvature as required by optimum aerodynamic shapes and for very appealing look (importance of esthetics).

Today's fibers (glass, nylon, Kevlar, carbon, whiskers or single crystal fibers of various chemical composition) are very strong, thus the structure becomes very light. The drawback is very little stiffness. The structure needs stiffening which is achieved either by the usual discreet stiffeners, or more elegantly with a *sandwich* structure: two layers of thin uni- or two-directional fibers are held apart by a lightweight core (foam or "honeycomb"). This allows the designer to achieve the required inertia or stiffness.

From an engineering standpoint, this method is very attactive and supported by many authorities because it allows new developments which are required in case of war. (The U.S. having no titanium or chromium needs to develop practical alternatives.) But this method also has its drawbacks for homebuilding: A mold is needed, and very strict quality control is a must for the right amount of fibers and resin and for good adhesion between both to prevent too "dry" or "wet" a structure. Also the curing of the resin is quite sensitive to temperature, humidity and pressure. Finally, the resins are active chemicals which will not only produce the well-known allergies (which is an increasing "disease" of today's human due to our weakened bodies from eating "junk food") but also the chemicals attack our body (especially the eyes and lungs) and they have the unfortunate property of being cumulatively damaging and the result (in particular deterioration of the eye) shows up only years after initial contact.

Another disadvantage of the resins is the limited shelf life, i.e., if the resin is not used within the specified time lapse after manufacturing, the results may be unsatisfactory and unsafe.

Finally unless the molds are very well designed, manufactured and maintained, the outside of the structure needs an often underestimated amount of "elbow grease" to provide the desired finish. Also a lot of care must be exercised as sanding down too much will result in a weaker structure. Historically the composites had their peak a couple of years ago. Today it is known (and proven by all those homebuilder "workshops") that only specialists can come up with a reliable and perfect structure and even the specialists take a chance on their own health.

LET'S SUMMARIZE

— Nature provides a raw material beautifully suited to aircraft structures. Unfortunately we are exploiting nature and today it is hard to find supplies of wood and also plywood of the required sizes and quality.

— Aluminum alloys in extruded and laminated form are an attractive alternative especially as they are easy to support with guaranteed properties.

— Steel tubing continues to be very popular in North America as welding does not seem to create any problems as feared in other parts of the world. A tubular structure is fabric covered.

— Composites can be looked at as "artificial wood" from a structural standpoint. Like everything artificial, it can be better than the natural product but the manufacturer needs to incorporate in the manufacturing process the wisdom provided by nature and/or the quality provided by other raw material's manufacturers (aluminum, chrome, moly steel). This is on top of an expensive mold, and the hazards to our own health (and our family's health when building in the basement).

How To Control Fuselage And Landing Gear Alignment

by Antoni (Tony) Bingelis
EAA Technical Counselor, Chairman

If you want an airplane that tracks through the air like an arrow and not one that sort of sidles along like a scavenging land crab, you will have to control its alignment carefully during the construction and assembly phases.

A well-built structure is one whose alignment is inconspicious. If everything looks reasonably aligned nobody notices it as most of us are blissfully oblivious of the obvious norm. Let the alignment of the rudder, wing, or stabilizer end up leaning a bit from that imaginary reference line though, and you will find yourself cringing everytime you look in that direction and hear your share of snide remarks about it as a constant reminder to your transgressions.

While the requirement for good alignment during assembly must be considered as essential . . . aesthetically, it is always critical. Fortunately for some of us, absolute accuracy in alignment, although always most desirable, is not always a prerequisite to an acceptable aerodynamic performance. Sometimes, however, misalignment in a component can portend disastrous consequences.

If, for instance, the incidence of one wing is greater than the other . . . bad news. Likewise, if one wing is warped during the skinning process, poor flight characteristics will result. It is a matter of concern because even a slight twist in one wing can lead to control difficulties and undesirable stall behavior. A large error can even render the aircraft uncontrollable. Figures 1 and 2 illustrate some of the common boo-boos that any builder can make without even trying. (Notice the statement read **any builder,** and not just an inexperienced one!)

A misaligned fuselage can generate more problems than anyone deserves. This becomes apparent when we realize that the fuselage, in most designs, is the foundation for the rest of the airplane. Quite often the wings, tail surfaces, and even the landing gear attach directly to some part of the fuselage. The attachment of these structural components are often by means of metal fittings bolted or welded to the fuselage at a particular station or bulkhead. If a builder is not careful, he will assume that that bulkhead or station location automatically locates his fitting for him and that no further alignment measurements are necessary . . . how wrong.

It is easy to be lulled into a premature sense of well being after you have completed two identical fuselage sides in the same jig. How could anything go wrong with the assembly after that? Aren't both sides exactly alike and isn't it simply a matter of cutting and fitting the cross members? True, but if the symmetry of the fuselage and the proper placement of

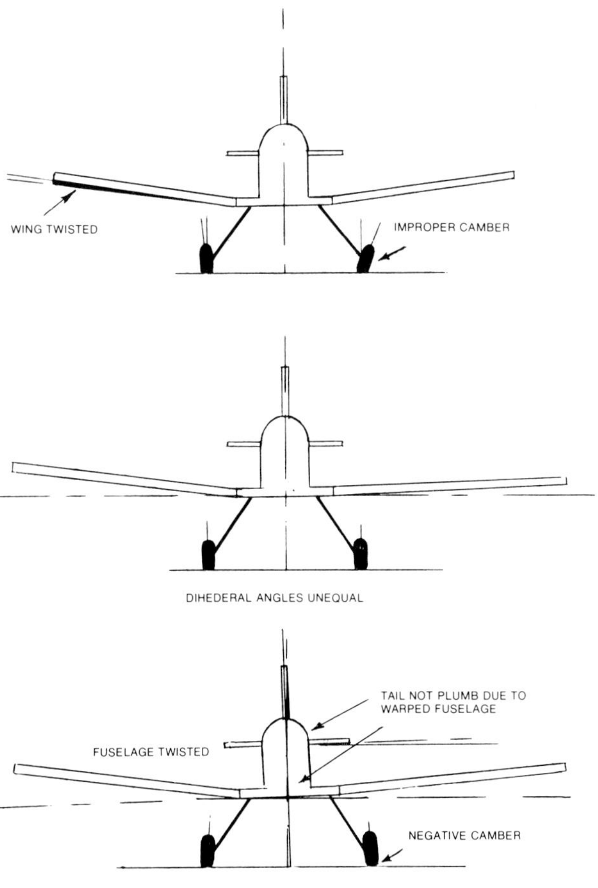

Figure 1

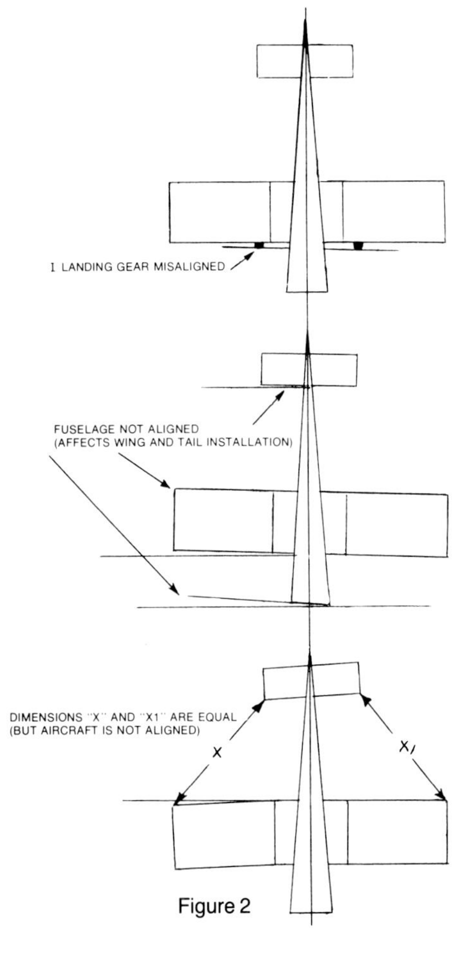

Figure 2

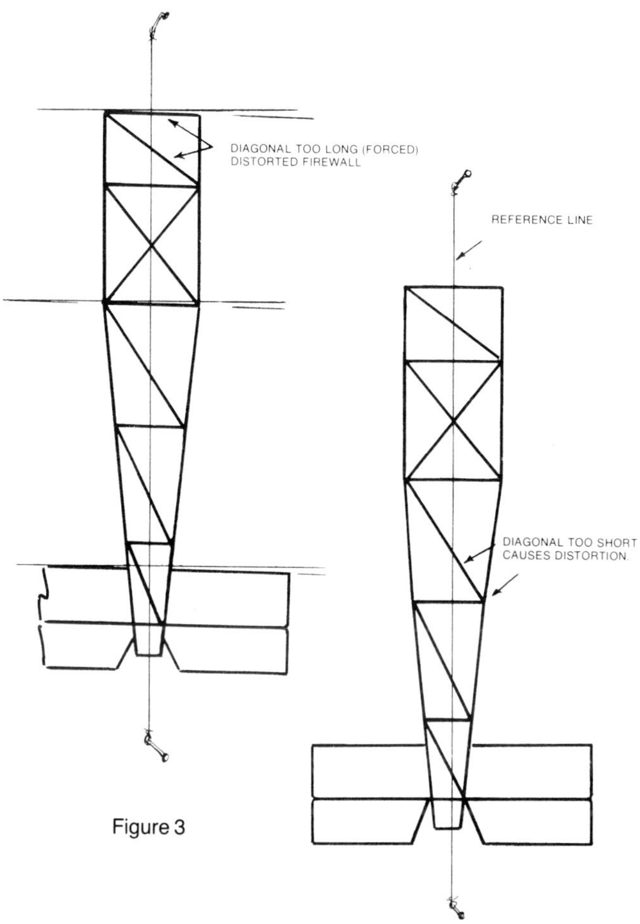

Figure 3

Figure 4

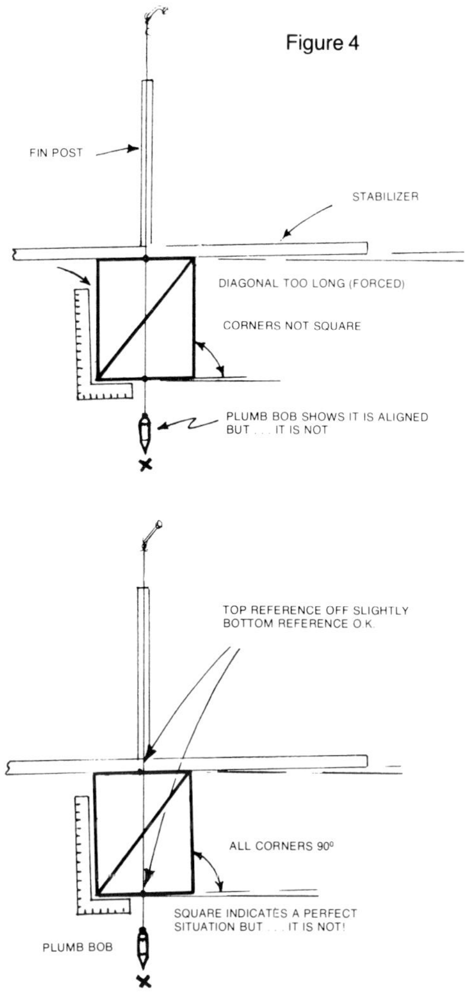

the firewall and the various cross members and bulkheads are not geometrically controlled, the whole fuselage can be thrown off alignment. If that happens, then your wing and tail surfaces and landing gear may go on cockeyed.

Before you start the assembly of the fuselage, a reference line should be established on your work surface over which the fuselage sides will be erected. This reference can be a drawn line or a tightly stretched wire or cord. An inked line drawn with a felt pen makes a very visible and effective reference. A stretched line can only be used if the fuselage is erected on raised supports that permits the presence of the line beneath without interference with the assembly. At one end, a line should be drawn across the reference line at a 90° angle. This is a very important starting point as the firewall will be positioned over it. If the firewall is not perfectly oriented with the reference line, any error at that point (station) will be locked in and will be continued all the way through the assembly of the rest of the fuselage. This can be serious if not caught in time and compensating corrections made to the engine mount, wing and tail attachment points. Another way the fuselage can be off is in its cross section. If the fuselage sides are not monitored continuously for squareness with a large carpenter's square, it is very possible that the cross section may assume the semblance of a parallelogram instead of maintaining the desired rectangular shape. Such a cross sectional warping would undoubtedly also have an adverse effect on the wing alignment, the fitting of the gas tank, and perhaps, the alignment of the instrument panel.

If the symmetrical and cross sectional integrity is not maintained, the rudder post and the stabilizer installation can become very difficult as their location and alignment is definitely influenced by the fuselage and its reference points.

In low wing all wood fuselages the alignment of the wing with reference to the fuselage is usually fixed by the location of a main bulkhead. If this bulkhead does not set square and at the proper vertical angle to the fuselage reference line, you may have difficulty not only with the alignment of the wing, but also in the establishment of the proper incidence. Incidence error can result in an abnormal nose-low or nose-high flight attitude in cruise. Either condition is undesirable due to increased drag.

When installing a wing center section in the fuselage, an alignment error as small as 1/8" on one side of the fuselage could be compounded to a 1 1/2" error at the tip of a 12' wing. It is best, therefore, to use the full length spar or the assembled wing in making those first alignment trial fits.

WHAT CAUSES FUSELAGE ALIGNMENT ERRORS?

(1) The most common cause is the builder's failure to establish and work from a centerline or reference line during the assembly of the fuselage. (It is also a common mistake to abandon reliance on the centerline too soon.)

(2) Failure to align the firewall at right angles to the centerline.

(3) Failure to mark the center of each station's cross member or bulkhead and then . . . failing to make sure, with a plumb bob, that this mark straddles the fuselage reference line.

(4) Permitting the structure to be moved accidentally . . . even a slight jar will change the reference points. Keep checking periodically.

(5) A glob of dried glue, a nut, a bolt, a file, or almost

Figure 5

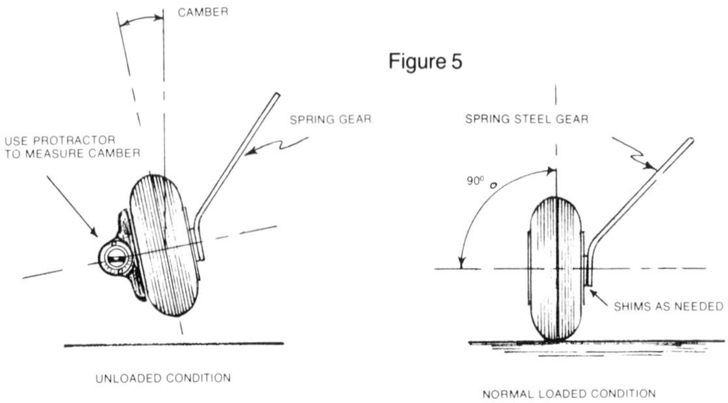

Figure 6

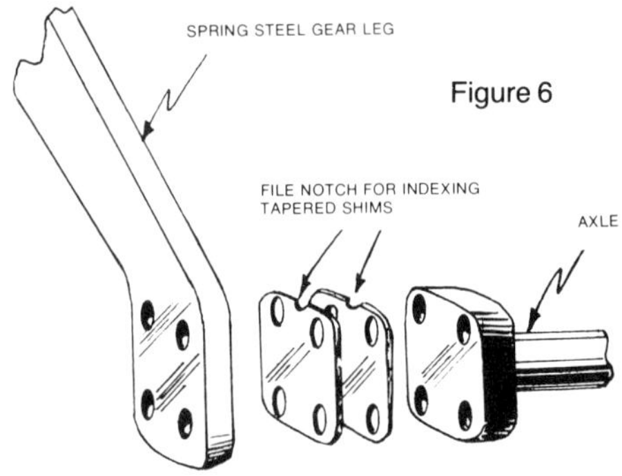

anything else lying on your work surface, can accidentally find itself under one of the longerons causing it to be raised slightly off the work surface. If you don't notice it, you may proceed with the assembly of the fuselage with one side of it slightly raised.

(6) Forcing any diagonal into place, however gently, will deflect the structure at that station and either the top or bottom cross members will no longer intersect the reference line. A too short diagonal member will have a similar effect.

There are three ways of doing your alignment checking. Visually (eyeball it), measure it with a steel tape, or use instruments such as a square, spirit level, combination square, protractor, and beam trammel. In practice, however, you will find it more practical to use any combination of the three methods to the degree best suited to the job at hand. Often a point is reached where the aspired for accuracy is more a fetish than a practical requirement. For example, to achieve the ultimate in accuracy, a guy can become frustrated trying to split a 1⁄16" difference in wheel alignment as measured at a point well ahead of the wheel's foot print.

Rectangular fuselages are checked at each station using a plumb bob and a large carpenter's square (See figure 2). When constructing any fuselage whose bottom longerons are closer together than the top longerons, the cross sectional check at each station must be accomplished using a plumb bob (see figure 3). An alternate method would be to use a combination protractor head to check the angle formed by each side of the fuselage. The angle reading for each side should be identical. In using this method, however, make sure that the fuselage is resting on a perfectly level surface.

Still another check should be made along the side profile when bulkheads are installed to insure vertical alignment. A bulkhead that leans forward or backward could, in some cases, alter the incidence angle when the wing installation is attempted. When fitting and installing the cross pieces in the typical welded steel tube or wood fuselage, each member should be centered over the reference line and fuselage sides checked for squareness using a large square. The greatest single source of fuselage misalignment occurs during the fitting of the diagonal members. If the fit of a diagonal member is a bit snug it will almost certainly deflect the fuselage from the reference line. Likewise, if a diagonal is just a bit too short and you have to use clamps to draw the fuselage sides in, that too will cause the fuselage to distort. As you work from the front of the fuselage toward the tail you will be tempted to prematurely abandon the checks because everything seems to be going fine . . . don't give in to the temptation. Keep close check on your reference line and each station until you get to the tail end.

LANDING GEAR AND WHEEL ALIGNMENT

Misalignment of the gear may not be very noticeable, but it can result in some mighty exciting runway performances during the high speed taxi tests and particularly before and after that important first flight. Some builders proclaim the wisdom of putting in a tiny bit of toe-in while some others say a dab of toe-out is best. Some like camber, others don't. Regardless which school of thought you subscribe to, the least you can do is to assure yourself that both wheels have the same amount of whatever it is you want.

Some landing gears and wheels cannot be adjusted for alignment after installation on the aircraft because the struts are attached to the fuselage at welded attach points, which cannot be adjusted. The axles, being welded to the struts, likewise cannot be adjusted. This type of gear simply needs to be properly aligned and jigged before welding. To accomplish this means that the airplane has to be raised to a level attitude and a reference or centerline established on the floor to provide guidance in jigging the gear for the alignment and welding.

Other landing gears using the vertical strut and scissors arrangement can often be adjusted to a limited degree by adding or removing washers from between the scissor elbow hinge (scissor elbow hinge!?)

The tapered steel whip rod gear is another that has to be jigged very accurately before the welding is attempted as it too, lacks a means of adjustment after installation.

The spring steel gear is a popular installation and it has the additional attraction of being ground adjustable. Of course, no matter what type of gear installation you use, every attempt should be made to insure accurate alignment at the very outset. At any rate you can easily adjust the toe-in and camber on a spring steel gear in accordance with design recommendations. In the absence of any such guidance, it is suggested that the best alignment is one that gives you a zero toe-in and a zero wheel camber at normal gross weight. The proper amount of camber is especially important with spring steel gears.

If your aircraft is larger than a single place job but you normally fly alone, and at less than gross weight, you might want to adjust your wheel camber to give a zero angle reading at this lesser operating weight to reduce wear along the outer edge of the tires.

To make these adjustments tapered metal shims are used to obtain the exact alignment and camber desired. After you have driven yourself batty trying different combinations of shims to get the wheels aligned the way you want them, you may be receptive to a suggestion. Index or mark the shims, and installation in exactly the same position will be guaranteed. Make your index marks by using a small rat tail file to cut a small groove across the top of all the shims. File the notch off-center so that the shims cannot be inadvertently reversed in assembly. Use one notch for the shims used in the left gear and two notches across the shims for the right gear.

Any time in the future that it may be necessary to remove the axle from the gear leg, the shims will already be marked assuring rapid and positive reassembly. This, of course, will guarantee that your wheel alignment is not disturbed any time you have to remove a wheel and axle for maintenance or inspection.

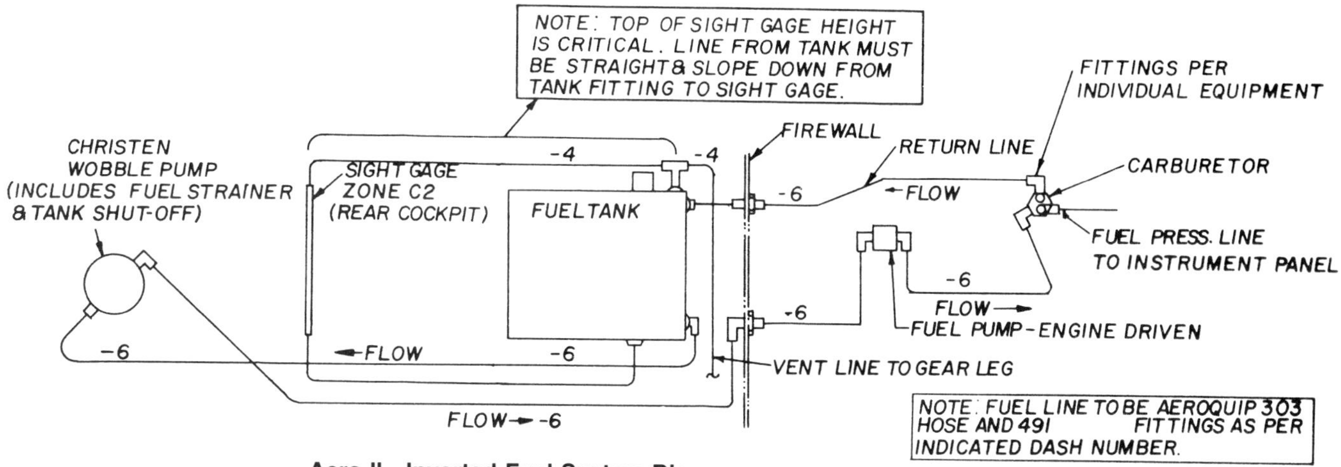

Acro II - Inverted Fuel System Diagram

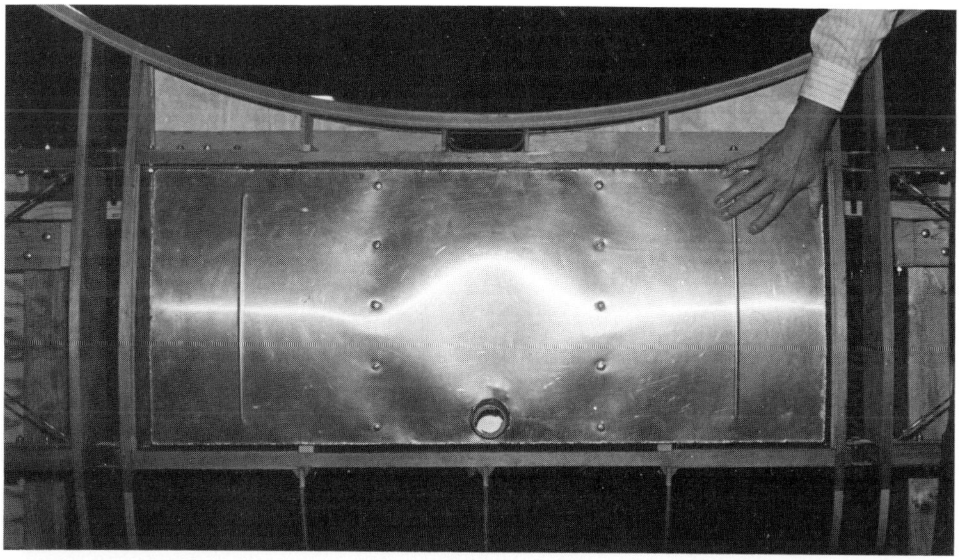

Fitting the center section wing tank in the Acro II, prior to plywood skin being glued to top surface of the wing. Tank holds 14.5 gallons of fuel. Add that to the 24 gallon capacity fuselage tank and the Acro II has respectable cross country legs. Good cruise speed gets you there fast, too!

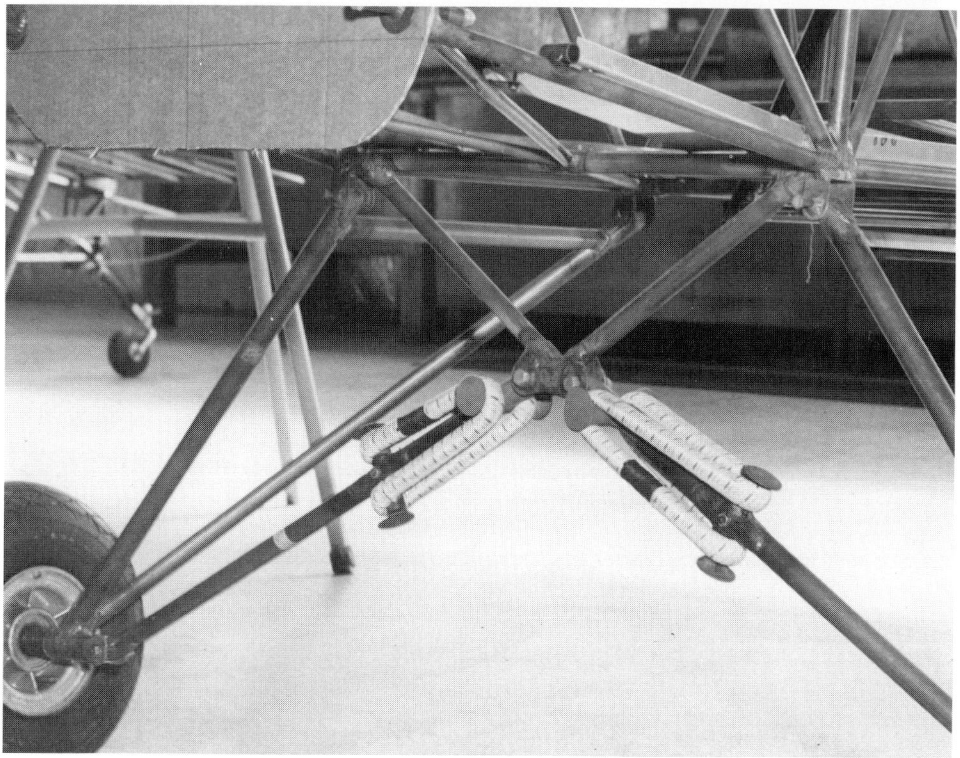

How shock cords are attached to the Cub type Pober Pixie gear. With a special tool, mounting shock cords becomes an easy task. This type installation makes replacement of cords very simple, and no disassembly is necessary as everything is in the open and easy to get at.

Fig. 1

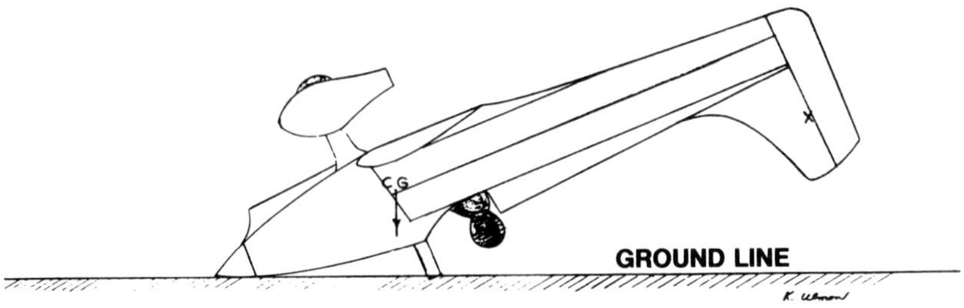

Fig. 2

Turnover Structures

By Kirk Ulman, EAA 10938

Federal air regulations state that the structure of an aircraft shall be designed to protect occupants from injury in event of a turnover unless the design of the aircraft generally precludes turning over. Most of us will interpret this to mean that an airplane with a conventional landing gear should have an overturn structure. Biplanes and high-wing airplanes generally have inherent overturn protection; therefore, we will consider low and mid-wing ships.

Crash helmets and shoulder harnesses are fine, but they will not keep an airplane from coming down on a pilot and breaking his neck. There are few of us who have not seen a light plane on its back. Most of the overturned ships have been high-wing jobs where the occupants have been suitably protected, but remember the damage the airplane sustained from a rather easy turnover; buckled struts, broken windshield, bent tubes, and a cowling smashed flat against the engine. Remember especially the buckled struts. Those struts are designed to take quite a negative flight load without buckling.

The old P-26 had an enormously high headrest. The first few models had normal size headreasts; but pilots were killed when the ships turned over, and the high headrest had to be provided for protection. This shows the fallacy in most cases of a turnover structure just an inch or two higher than the pilot's head located behind the cockpit.

Figures 1 and 2 depict an airplane of size and proportion similar to the typical homebuilt airplane. In Figure 1 the turnover structure is shown located behind the cockpit. Notice that the top of the pilot's head is actually below the ground line. This is protection? In most cases a turnover structure behind the cockpit will have to be very high and heavy to do any real good.

Figure 2 shows the airplane with a turnover pylon of same height located ahead of the pilot. Note the distance afforded between the pilot's head and the ground. Even though a turnover structure is located ahead of the cockpit, it is desirable in many cases to have it slightly aft of the center of gravity as shown in Figure 2 so that the normal inverted resting position will allow more room for exit between the airplane and the ground. If the aircraft momentarily rocked back on the tail when turning over and the fin were to be smashed to nearly one-half its original height, we can visualize the clearance afforded by projecting a line from the top of the turnover structure to this height point on the tail (X mark). We can do this to the airplane in Figure 2 and see that there is still pilot protection.

On an existing airplane, turnover protection can be determined by laying a board or length of electrical conduit on top of the airplane from the engine to the highest piece of solid structure. Don't deceive yourself either, as a flimsy windshield frame isn't solid structure. A piece of string stretched from the engine to the highest piece of solid structure will also do. The important thing is to make sure the pilot's head does not protrude above this projected line.

Once we have determined where the turnover structure should be located, we are faced with the slightly more vexing problem of how strong to make it. The conditions of the turnover greatly affect the impact force on the cockpit area, and the configuration of the aircraft is also a factor. It would be ridiculous to try to stress the structure to withstand the impact force of a dead stop turnover at 70 mph. The high translational velocity becomes a rotational velocity which diminishes to zero in a very short time at ground contact. Fortunately, the chances of a dead stop, high speed turnover are remote as the aircraft will generally slide on its back or tumble again; and in so doing will dissipate energy more gradually. The seat belt helps us determine the necessary strength of the turnover structure. Why have the extra weight of a 20G turnover structure when the seat belt separates at 7Gs and throws the pilot out on his head?

The law requires a single seat belt withstand only a 1,000 lb. load; but in my experience of static testing seat belts, I have found that even a light belt in good condition will take a 7G load (170 lb. man) without complete failure. The conclusion we draw now is that a 7G turnover structure should do the job, but let us consider one more thing. We should use a factor of safety in the design of our turnover pylon, and anyway we might have a strong seat belt that will take more than 7Gs. FARs state a minimum ultimate factor of safety of 1.50 be used in computing passenger loads. This regulation is a good guide, and we will use it here. The turnover structure should withstand 7Gs × 1.50 = 10.5Gs. If an airplane has a gross weight of 900 lbs. (900 × 10.5 = 9,450).

If we put the turnover structure in front of the cockpit and do not wish to use the elaborate build-up military type pylon, we can build a reinforced windshield frame (Fig. 3). In using this method of construction it is desirable to keep the side legs of the frame straight and account for the curvature of the windshield in the uppermost portion because a member will tend to buckle where it is already bent. An even simpler arrangement can be adapted to many strut-braced low-wing ships where the cross member of the fuselage at the strut

Fig. 3

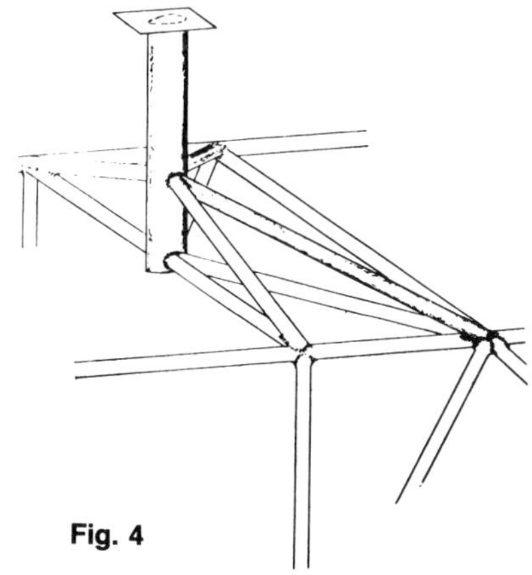

Fig. 4

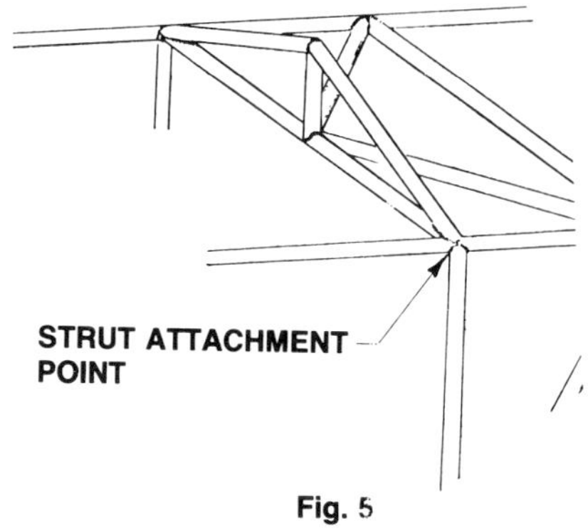

Fig. 5

attachment is reinforced against buckling as in Figure 4. Figure 5 shows the structure slightly modified with the addition of two lateral brace tubes and single tube pylon.

One is not limited to the use of steel tubes in constructing an overturn structure, but the strength of steel tubes in compression for various lengths can be found in many aircraft reference books if one is not too sharp on his column formulas, and it might be easier to get a more accurate estimate of the structure's strength.

These features should be incorporated in the design of all turnover structures: The top of the structure should have enough area to prevent surface penetration on impact. The structure should tie in with tube clusters or other suitable reinforced parts of the fuselage, and the structure should be braced against side loads which could be encouraged during an inverted slide. This may call for an oversize tube on the single tube pylon structure.

While we are on this unpleasant subject of turnover, let us once again go into the rule book. It says that the pilot must be able to exit from the aircraft easily if it turns over. It would not be nice to be trapped inside if the airplane caught on fire. We see many cardinal violations of this rule. The tilt forward, tilt aft, side hinged, and lift-off hatches could easily turn a low-wing conventionally geared aircraft into a pyre. A slide open hatch which can be left open for ground operations and take-offs and landings is best safety-wise. Also, an open cockpit should have sides low enough to permit easy egress.

The next time I see a guy with poor turnover protection on his airplane, I will tell his wife; and, if she is anything like mine, she will make him fix it.

Designer of the parasol Pixie, Paul Poberezny, places a completed wing rib in position on the cabane struts. Cabane struts in designs such as this serve same purpose as turn over structure, protecting pilot in case of a nose over. Little sport plane is powered with a VW engine.

Crashworthiness and the Shoulder Harness

by Antoni (Tony) Bingelis
EAA Technical Counselor, Chairman

At an altitude of 200' soon after take-off, the engine lost power. The aircraft was observed gliding in a shallow left turn until it disappeared from view behind a knoll. The biplane impacted in a level attitude in a soybean field hard enough to fail both main gears and then slid to a stop in approximately 45 to 50'.

The seat belt held but . . . with no shoulder harness to restrain him, the pilot's head struck the top of the instrument panel displacing it forward.

An extruded aluminum section, illustrated in Figure 1, used to support the top of the instrument panel and secure the coaming pierced the pilot's head causing a fatal injury.

If this extruded aluminum section had not been in this location, the pilot probably would have ended up with nothing worse than a headache.

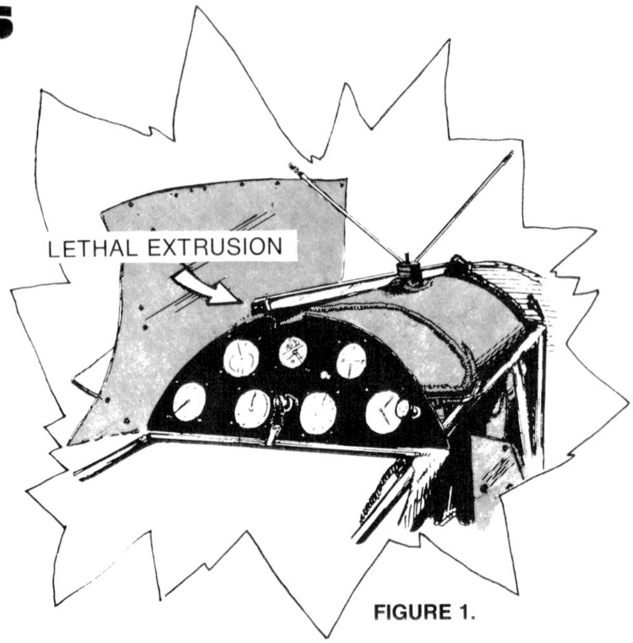

FIGURE 1.

IMMEDIATE ACTION IS RECOMMENDED

There was no reason for the spear-like extrusion or angle reinforcement to be positioned as shown in the drawing. If such a reinforcement is installed in your aircraft, it should be removed, or at least modified so that it does not extend beyond the fuel tank. The evidence shown in Figure 1, should be sufficient. This kind of reinforcement is more likely to be found in biplanes as presence of center section brace wires, in some designs, makes the installation of the instrument panel hood in one piece very difficult, if not impossible. Nevertheless, I would suggest that perhaps no reinforcement at all is needed as this is not a primary structure type of thing. That metal hood or coaming merely has a fairing and streamlining function in most cases. By using nut plates even a two piece coaming could be installed with no need for a reinforcement strip.

Equally as dangerous as the absence of a shoulder harness installation is the presence of protruding knobs, switches and ignition keys in the central portion of the instrument panel of single seaters. In the case of two seater aircraft, you should avoid the placement of such protruding devices in the areas directly in front of either the pilot or the passenger.

REMEMBER THE SHOULDER HARNESS PLEDGE?

The critical need for the installation and use of pilot and passenger restraint systems in sport aircraft was recognized and pioneered by the EAA for more than 20 years. EAA membership applications have long contained the Shoulder Harness Pledge. Do you remember how it goes? Most of us have signed it.

"I __________________ hereby promise to install and wear shoulder harness and safety belts in my private built aircraft to protect myself, passenger and the good name of the association. Air Force and Navy tests have proved that a 20G harness will eliminate 90% of aircraft accident injuries."

The necessity for pilot and passenger restraints is well documented with numerous examples of serious or fatal injuries sustained in potentially survivable accidents. By potentially survivable accidents we mean those in which the surrounding cockpit or cabin structure remains intact without serious deformations and where the level of impact forces is below expected human tolerances.

Time and again it is proved that one of the prime factors contributing to head injuries is due to the widespread reliance on only a seat belt to restrain the body. A seat belt alone cannot provide the needed protection.

The installation and proper use of a shoulder harness will, however, help eliminate the jackknifing action of the body in event of sudden exposure to abrupt impact forces. See Figures 2 and 3.

SELECT ATTACHMENT POINTS CAREFULLY

Each builder should make a critical analysis of his aircraft to establish the best location for the seat belt and shoulder harness attachment points. If the seat belt is to be attached to the seat the structural integrity should carry through to the primary aircraft structure.

If the aircraft is being built from plans, the seat belt attachments and shoulder harness attachment points will probably have been worked out and detailed on the plans. It is too bad, though that a few of the aircraft designs have what I

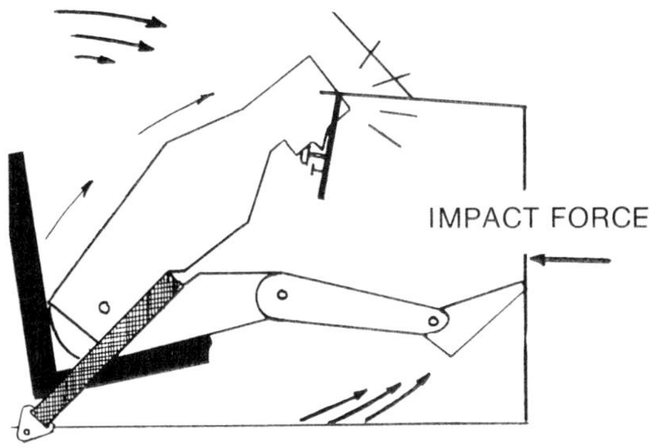

ELIMINATE JACKKNIFING EFFECT
FIGURE 2.

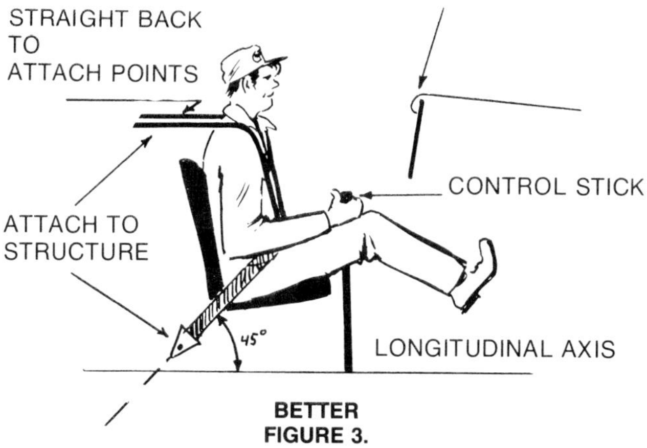

BETTER
FIGURE 3.

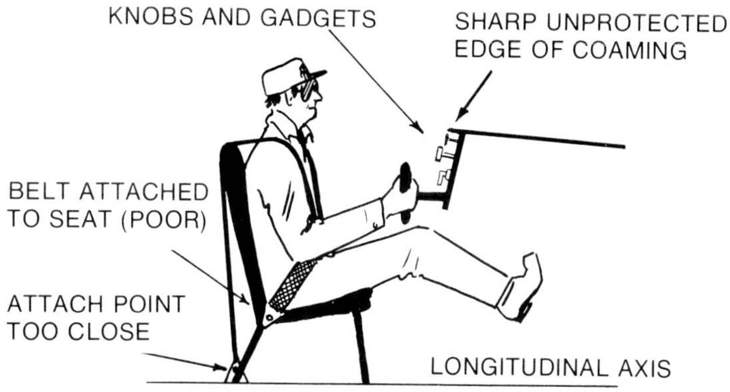

POOR
FIGURE 4.

consider to be rather inferior shoulder harness installations . . . my own Emeraude, for example.

This basic fault is as common to all-wood aircraft as it is to the tube and rag types. The particular deficiency I have in mind is that attributed to the use of the unreinforced seat back or fuselage cross member as an attachment or loop-over point for the shoulder harness. This method does provide some protection . . . but not much. Such structural members will fail long before the belts themselves fail. The completed installation might give you a nice moral feeling but very little protection. Aircraft with such installations should perhaps have a section of 1/8" flexible control cable running aft to some solid point on the aircraft as an additional restraining point. This would serve to increase the strength of the restraint system considerably.

The builder should also be careful not to weaken the basic structure of his aircraft by drilling holes through unreinforced portions of his longerons just to find a good place for the belt and harness attachments.

Not only must the points of attachment be strong enough to develop the full potential of the restraining belts, but the attachment ends of the belts must present and maintain certain relative positions or angles as they pass over the lap and shoulders.

The shoulder harness must also be so positioned as to permit limited freedom to accommodate normal body movements of the pilot and passenger without head-neck contact or interference with vision.

Typical good and poor restraint configurations are illustrated in Figures 3 and 4. Do you see the basic differences?

The installation of shoulder harnesses is really uncomplicated and at worst would require but little change or modification of the structure to accommodate them.

Undoubtedly the four place aircraft causes the greatest difficulty when it comes to finding a suitable attachment location for the shoulder harness belts. Part of this is due to the likelihood that the front seat harness might unduly restrict the vision of the rear seat passengers. This sort of inconvenience can be minimized, however, by making use of suitable attachment points located in the upper-side areas of the fuselage structure.

A GOOD INSTALLATION CAN STILL BE BAD

Alright, so your seat belt and shoulder harness installation is a pretty good one and should be effective. Well, if you are over that hurdle this might also be a good time to think about the physical condition of your belts and harnesses. Were they new when you installed them? What sort of condition are they in now? The sun is pretty rough on most material, especially fabrics. Take a good look, for after all when sudden stoppage of an aircraft occurs, the difference between little or no injury and disaster often hinges on the condition of the belts and their attachments to the aircraft structure.

Replacement of your seat belts and harnesses is recommended when:

1. Frayed belt edges have become quite apparent to you.
2. There are indications that the stitching is deteriorating.
3. The buckle serrations have become worn to the point where slippage is possible. (belts of this type)
4. The belt shows signs of deterioration.

THINK ABOUT IT . . .

It is natural enough for many builders to over-rate the protective benefits of the safety belt and as a result under-rate the value of the shoulder harness and the need for building crashworthiness into the cockpit and cabin installations. I'd say these gents are dead wrong and no pun is intended. I still recall early military training reports which the Air Force used to convince us of the value of these safety precautions. Later, studies by the Cornell University Medical College in New York, the FAA and others, all tended to strengthen the case for adequate body restraints and improved cockpit design to reduce bodily injury in survivable accidents.

All this is not new . . . in a report issued twenty years ago, the Cornell Medical College study group found that almost 80% of the survivors of light aircraft accidents, in which seat belts were the only body restraints used, sustained head injuries.

Any time you have the situation where the head is injured three times as often as the other areas of the body which are adjacent to the safety belt, it is mighty hard not to believe that some sort of restraints should be provided for the upper torso. The simplest means for this sort of protection, we realize, is a shoulder harness. Not just any shoulder harness but one that is sufficiently strong, properly anchored, and used. Take another look at Figure 1. It also makes a good case for the use of crash helmets, doesn't it?

EXPERIMENTAL AIRCRAFT ASSOCIATION ®

Weight And Balance

For The Aircraft Builder

by Dick Gleason

The weight and balance computations for an aircraft are by no means difficult nor are they something to be afraid of. The aircraft mechanic must know and understand the principles of airplane weight and balance to properly balance aircraft for safe flight. The pilot-owner should know the rudiments of weight and balance to insure proper loading for safe flight. Load distribution is important for safe flight, and if not properly done may result in hazardous instability.

Weight and balance does not require extensive study — once the basic principles and computation are learned, it becomes an interesting subject. Certain combinations of nose and tail loads are critical in flight stability, therefore it is necessary that precautions be taken to obtain the proper location of the aircraft's center of gravity on the longitudinal axis.

The following definitions are common to all methods of weight and balance control and should be studied. Understanding these terms clarifies a lot in the operation of aircraft.

Maximum Gross Weight (MGW): The maximum authorized weight of the aircraft and its contents as listed in the particular aircraft specification or as computed in the design of the aircraft.

Empty Weight (EW): The unloaded weight which includes all operating equipment that has a fixed location and is actually installed in the aircraft, such as batteries, radios, floats, lights, etc.

Useful Load (UL): The useful load is the difference between the empty weight and the gross weight of the aircraft. It consists of the pilot, maximum oil, fuel, passengers, baggage, cargo, etc., unless otherwise noted. The useful load consists of both the pay and non-pay loads.

Aircraft Weight Check: This consists of checking the sum of the weight of all items of useful load against the authorized useful load (max. weight less empty weight) of the aircraft.

Datum (D): An imaginary vertical plane or line from which all horizontal measurements are taken for balance purposes, with the aircraft in LEVEL FLIGHT ATTITUDE.The datum may be at the wing leading edge, nose of the aircraft, front face of the firewall, a position in front of the nose, at the main spar, or even the rear edge of the cockpit. If not specified, the datum may be located in any convenient position on the aircraft so long as it is indicated and so recorded. (See Fig. 1)

Arm (or Moment Arm): The horizontal distance, in inches, from the datum to the center of gravity of an item. A plus (+) arm indicates the item is located to the rear of the datum. A minus (−) arm indicates the item is located forward of the datum. (See Fig. 2)

Moment: The moment of an item about the datum is obtained by multiplying the weight of the item by its horizontal distance from the datum. (See Fig. 2)

Center of Gravity (CG): Generally known as the balance point of an aircraft. It is a point about which the nose-heavy and tail-heavy moments are exactly equal in magnitude. (See Fig. 2a)

Empty Weight Center of Gravity (EWCG): The center of gravity of an aircraft in its empty weight condition. (See Fig. 10)

Operating Center of Gravity Range: The distance between the forward and rearward CG limits shown in the aircraft's specifications or as determined by design computations. (See Fig. 3)

Mean Aerodynamic Chord (MAC): The mean chord of the wing. For weight and balance purposes, it is used to locate the CG range of the aircraft. MAC data will be found in the aircraft specifications, flight manual, or aircraft weight and balance records, or determined by design computations.

Weighing Point: With the aircraft in level flight attitude on the scales, a vertical line through the wheel centers will indicate the point at which the weight is concentrated. The horizontal fore and aft distance between the points on the scales is noted for later use in solving the weight and balance problem. (See Fig. 4)

Minimum Fuel: Minimum fuel weight is necessary in computing the most forward CG limit; for this purpose 1/12th gallon per METO (max. - except take-off horsepower) is customary. To determine the weight of fuel in pounds, divide the METO hp by 12. Fuel weighs 6 lbs. per U. S. gallon.

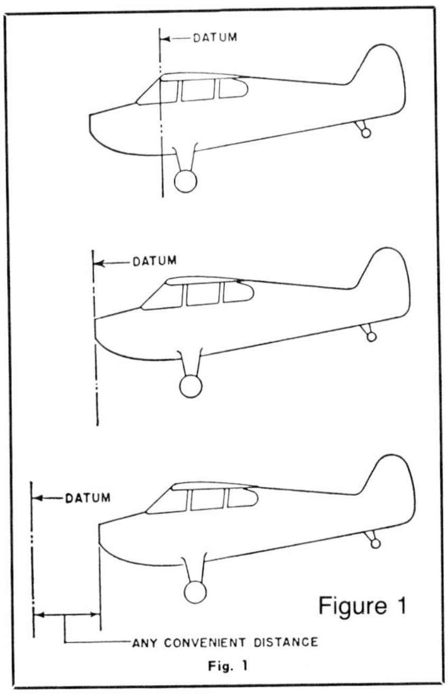

Figure 1

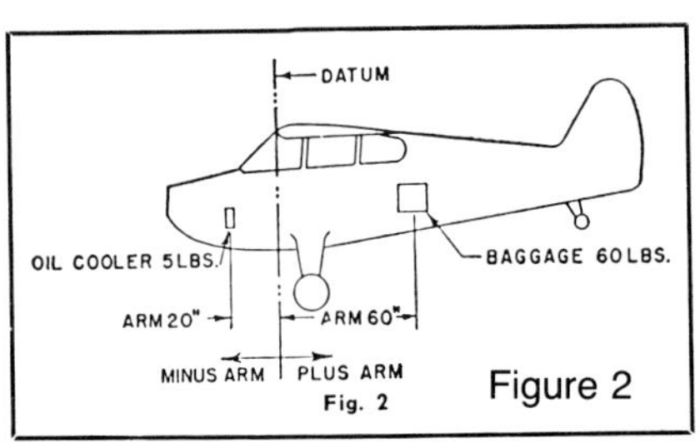

Figure 2

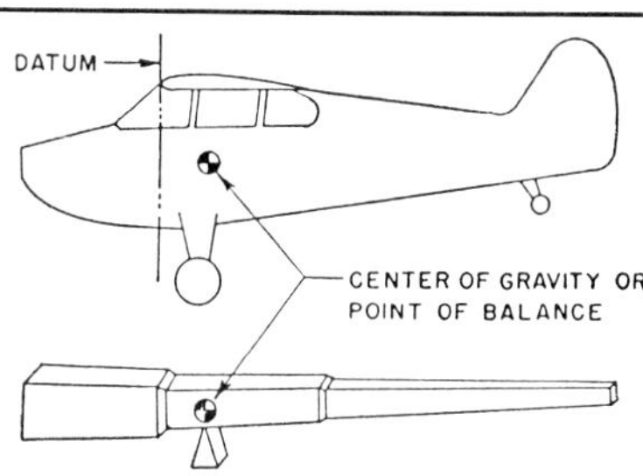

The entire aircraft weight may be considered to be concentrated at the center of gravity. Therefore, the moment of the aircraft about the datum is the weight of the aircraft times the horizontal distance between the C.G. and the datum.

Example: If the weight of this airplane is 2000 lbs. and the arm from the datum to the center of gravity is 16 inches, the moment of the aircraft about the datum is 2000 x 16 or 32,000 in. lbs.

Figure 2A

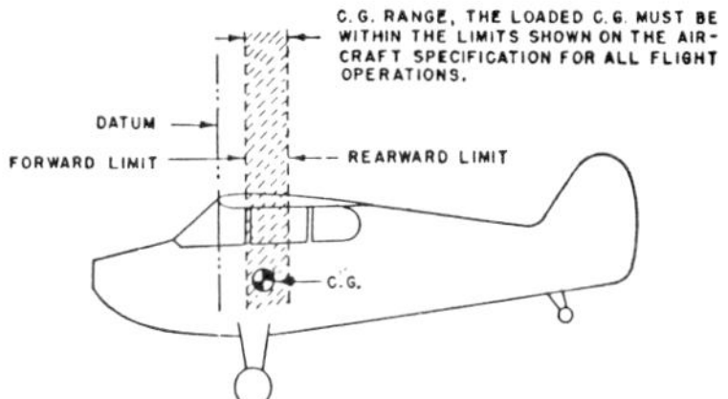

Figure 3

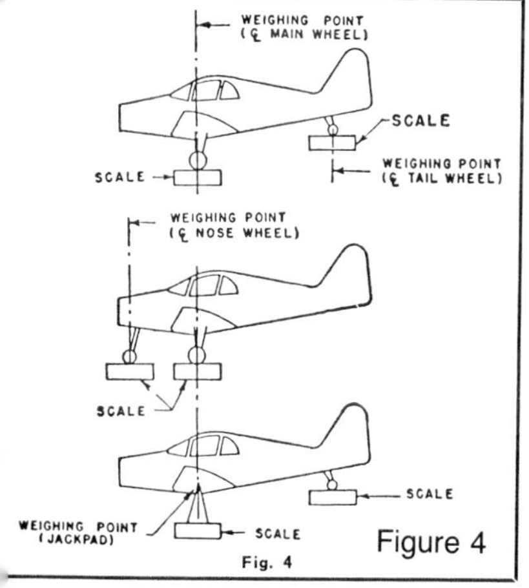

Fig. 4

Figure 4

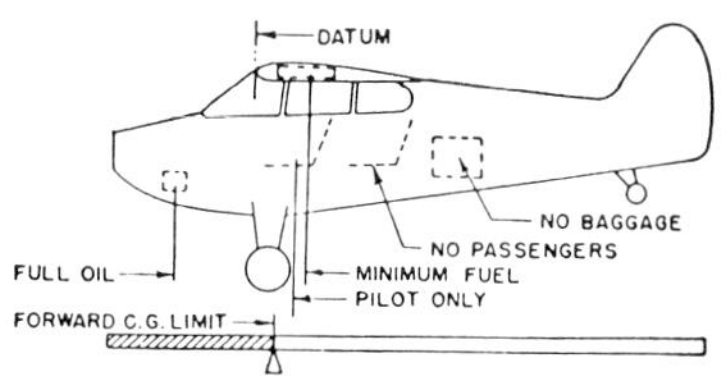

TO CHECK: MOST FORWARD WEIGHT AND BALANCE EXTREME.

GIVEN: Actual empty weight of the airplane -------------- 1169#
Empty weight center of gravity ---------------- +10.6''
*Maximum weight -------------------------- 2100#
*Forward C.G. limit ------------------------- + 8.5''
*Oil, capacity 9 qts. -------------------------- 17# at - 49
*Pilot in farthest forward seat equipped with controls (unless otherwise placarded) ---------- 170# at + 16''
*Since the fuel tank is located to the rear of the forward C.G. limit, minimum fuel should be included. $\frac{\text{METO HP}}{12} = \frac{165}{12} =$ 13.75 gal. x 6# ------- 83# at + 22''

*Information should be obtained from the aircraft specification.

Note: Any items or passengers must be used if they are located ahead of the forward C.G. limit.
Full fuel must be used if the tank is located ahead of the forward C.G. limit.

CHECK OF FORWARD WEIGHT AND BALANCE EXTREME

	Weight (#)	x Arm ('')	Moment (''#)
Aircraft Empty	+ 1169	+ 10.6	+ 12391
Oil	+ 17	- 49	- 833
Pilot	+ 170	+ 16	+ 2720
Fuel	+ 83	+ 22	+ 1826
Total	+ 1439 (TW)		+ 16104 (TM)

Divide the TM (Total Moment) by the TW (Total Weight) to obtain the forward weight and balance extreme.

$\frac{TM}{TW} = \frac{16104}{1439} = +$ 11.2''

Since the forward C.G. limit and the maximum weight are not exceeded, the forward weight and balance extreme condition is satisfactory.

Figure 5

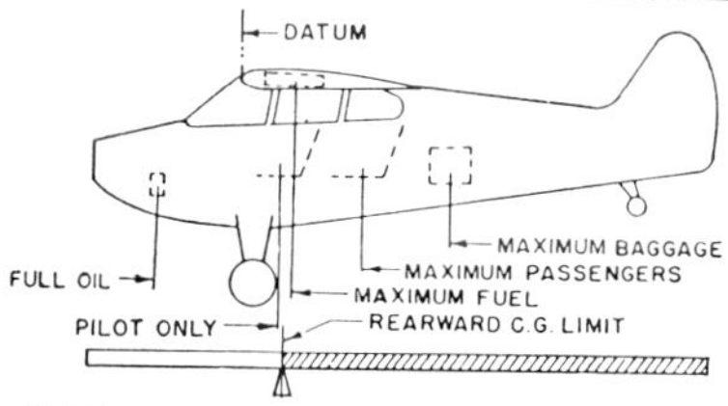

TO CHECK: MOST REARWARD WEIGHT AND BALANCE EXTREME.

GIVEN: Actual empty weight of the airplane --------- 1169#
Empty weight center of gravity ------------ 10.6''
*Maximum weight ---------------------- 2100#
*Rearward C.G. limit -------------------- 21.9''
*Oil capacity 9 qts. --------------------- 17# at - 49''
*Baggage, placarded do not exceed 100 lbs. --- 100# at + 75.5''
*Two passengers in rear seats, 170 x 2 ------ 340# at + 48''
*Pilot in most rearward seat equipped with controls (unless otherwise placarded) ------- 170# at + 16''
*Since the fuel tank is located aft of the rearward C.G. limit full fuel must be used --- 240# at + 22''

* Information should be obtained from the aircraft specification.

Note: If fuel tanks are located ahead of the rearward C.G. limit minimum fuel should be used.

CHECK OF REARWARD WEIGHT AND BALANCE EXTREME

	Weight (#)	x Arm ('')	= Moment (''#)
Aircraft empty	+ 1169	+ 10.6	+ 12391
Oil	+ 17	- 49	- 833
Pilot (1)	+ 170	+ 16	+ 2720
Passengers (2)	+ 340	+ 48	+ 16320
Fuel (40 gals.)	+ 240	+ 22	+ 5280
Baggage	+ 100	+ 75.5	+ 7550
Total	+ 2036 (TW)		+ 43428 (TM)

Divide the TM (Total Moment) by the TW (Total Weight) to obtain the rearward weight and balance extreme.

$\frac{TM}{TW} = \frac{43428}{2036} = +$ 21.3''

Since the rearward C.G. limit and the maximum weight are not exceeded, the rearward weight and balance extreme condition is satisfactory.

Figure 6

MAKE MA-700 MODEL A SERIAL # 0000 REGISTRATION # N1234
DATUM IS loading edge of wing.

COMPUTE AS FOLLOWS IF AIRCRAFT WEIGHED

1. Leveling means: level top longeron between front and rear seats.
2. Main wheel weighing point is located (______"FORWARD)(+ 3 " AFT) of datum.
3. Actual measured distance from the main weight point centerline to the tail (or nose) point centerline 222 ".
4. Oil over and above "ZERO" tank reading = (a. — Gals.)(b. — Lbs.)(c. — In.)

ACTUAL EMPTY WEIGHT

	Weight Point	Scale Reading	- Tare	= Net Weight
5.	Right	564	0	564
6.	Left	565	0	565
7.	Tail	67	27	40
8.	Nose	—	—	—
9.	Total Net Weight			1169

CENTER OF GRAVITY AS WEIGHED

10. C.G. relative to main wheel weighing point:
(a) Tail wheel airc. (Item 3, 222) X (Item 7, 40) / (Item 9, 1169) = + 7.6 = C.G.
(b) Nose wheel airc. (Item 3 —) X (Item 8 —) / (Item 9 —) = ____ = C.G.

11. C.G. relative to datum:
(a) Tail wheel airc. (Item 10a, + 7.6) added to (Item 2, + 3) = + 10.6" = C.G.
(b) Nose wheel airc. (Item 10b, ____) added to (Item 2, ____) = ____ = C.G.

COMPUTE IF AIRCRAFT WEIGHED WITH OIL (Item 4)

	Weight	X Arm	= Moment
Aircraft	(9)	(11)	
Less Oil	(4b)	(4c)	
Empty Totals	(a)		(b)

12. (b)/(a) -------- = (c)--------" = Empty weight C.G.

REPAIR AGENCY ________ ________ DATE ________
Name Number

Fig. 7

EQUIPMENT LIST

*Required or Optional Item Numbers as Shown in Aircraft Specification						
1	2	101	102	103	104	105
106	201	202	203	301	302(a)	303
401(a)	402	—	—	—	—	—

Special Equipment

Item	Make	Model	Weight	Arm
3 Flares 1½ Min.	XYZ	03	25#	150"
Enter above those items included in the empty weight.				

WEIGHT AND BALANCE EXTREME CONDITIONS

Approved fwd. limit 8.5 " Approved max. weight 2100# Approved aft limit 21.9"

Item	FORWARD CHECK Weight	X Arm	= Moment	REARWARD CHECK Weight	X Arm	= Moment
Airc. Empty	+ 1169 (9 or 12a)	+ 10.6 (11 or 12c)	+ 12391	+ 1169 (9 or 12a)	+ 10.6 (11 or 12c)	+ 12391
Oil	+ 17	- 49	- 833	+ 17	- 49	- 833
Pilot	+ 170	+ 16	+ 2720	+ 170	+ 16	+ 2720
Fuel	+ 83	+ 22	+ 1826	+ 240	+ 22	+ 5280
Passenger(s)				+ 340	+ 48	+ 16320
Baggage				+ 100	+ 75.5	+ 7550
TOTAL	+ 1439 = TW		+ 16104 = TM	+ 2036 = TW		+ 43428 = TM

$\frac{TM}{TW} = \frac{16104}{1439} = +11.2" =$ Most Forward C.G. location

$\frac{TM}{TW} = \frac{43428}{2036} = +21.3" =$ Most rearward C.G. location

LOADING SCHEDULE

Gallons of Fuel	Number of Passengers	Pounds of Baggage
40	2(R)	100
The above includes pilot and capacity oil.		

EQUIPMENT CHANGE

Computing New C.G.

Item, Make, and Model*	Weight	X Arm	= Moment
Airc. Empty	+ 1169 (9 or 12a)	+ 10.6 (11 or 12c)	+ 12391
204 added	+ 6	- 1	- 6
302(b) added	+ 29	+ 13	+ 377
302(a) removed	- 24	- 29	+ 696
303 removed	- 1	+ 4	- 4
NET TOTALS	- 1179 = NW		+ 13454 = NM

$\frac{NM}{NW} = \frac{13454}{1179} = +$ 11.4" = New C.G.

*ITEM NUMBERS WHEN LISTED IN PERTINENT AIRCRAFT SPECIFICATION MAY BE USED IN LIEU OF "ITEM, MAKE, AND MODEL".

APPROVED BY ____________ DATE ____________

Fig. 8

Full Oil: The quantity of oil shown in the aircraft specifications as its oil capacity, or as shown on the engine oil supply tank or sump.

Tare: The weight of block, chocks, etc., used to secure the aircraft on the scales. Generally included in the initial scale readings, and must be deducted from the scale reading to arrive at the net weight of the aircraft.

Unit Weights: The following items are used for all weight and balance computations:

FUEL — six pounds per U.S. gallon.

OIL — seven and one-half pounds per U.S. gallon.

WATER — eight pounds per U.S. gallon.

CREW & PASS. — one hundred seventy pounds per person.

PARACHUTE — twenty pounds.

Algebraic Sign (±): Strict attention must be paid to the plus and minus signs in weight and balance computations, as care must be exercised to retain the proper sign in all

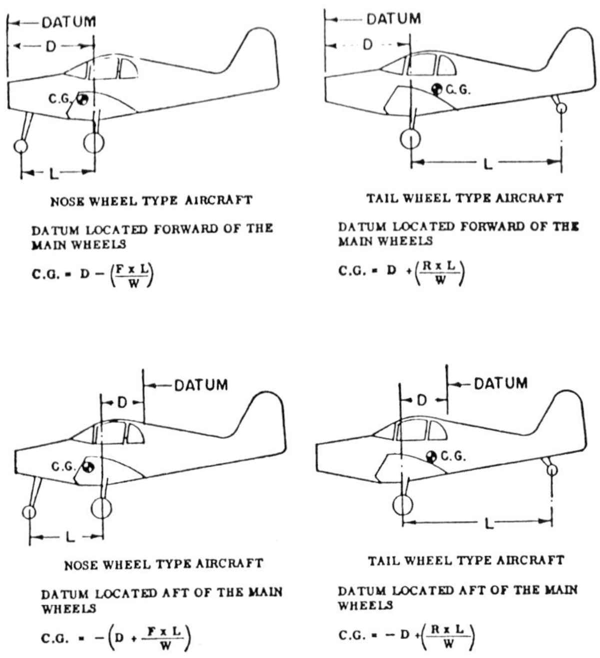

CG = Distance from datum to center of gravity of the aircraft.
W = The weight of the aircraft at the time of weighing.
D = The horizontal distance measured from the datum to the main wheel weighing point.
L = The horizontal distance measured from the main wheel weighing point to the nose or tail weighing point.
F = The weight at the nose weighing point.
R = The weight at the tail weighing point.

Figure 9

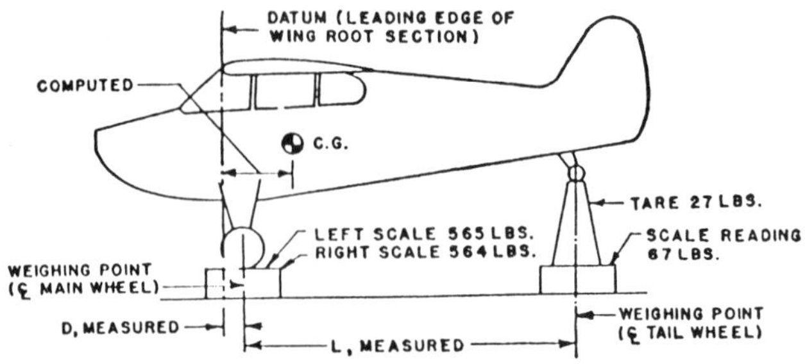

TO FIND: EMPTY WEIGHT AND EMPTY WEIGHT CENTER OF GRAVITY

Datum is the leading edge of the wing (from aircraft specification)
(D) Actual measured horizontal distance from the main wheel weighing point ℄ main wheel) to the Datum------- --3"
(L) Actual measured horizontal distance from the rear wheel weighing point (℄ rear wheel) to the main wheel weighing point------------------------------------222"

SOLVING: EMPTY WEIGHT

Weighing Point	Scale Reading #	Tare #	Net Weight #
Right	564	0	564
Left	565	0	565
Rear	67	27	40
Empty Weight (W)			1169

SOLVING: EMPTY WEIGHT CENTER OF GRAVITY

Formula: $C.G. = D + \frac{R \times L}{W} = 3'' + \frac{40 \times 222}{1169} = 3'' + 7.6'' = 10.6''$

Reference for formula Fig. 15-4

This case is shown properly entered on a sample weight and balance report form, Figure 15-17.

Figure 10

answers. Remember that the problem is solved as if the aircraft were being viewed from its left side with its nose to your left. (See Fig. 2)

Determination of Algebraic Signs (+ or −): Any weight item added to the aircraft, whether fore or aft of the datum, is a plus weight. Any weight item removed from the aircraft, whether fore or aft of the datum, is a minus weight. The arm of any item forward of the datum is a minus arm. The arm of any item aft of the datum is a plus arm. (See Fig. 2)

Weight and Balance Extreme Conditions: The most forward and rearward CG position for the aircraft. This position is determined by two separate computations. (See Figs. 5 and 6)

Equipment List: A complete list of all equipment installed in the aircraft and included in the certificated empty weight. Found in either the approved airplane operating manual, or the weight and balance data. (See Fig. 8)

Equipment Change: The owner of an aircraft is responsible for a continuous balance record of each aircraft, listing all changes affecting the weight, CG location, and equipment, in order that a computed weight and CG location may be established at any time. (See Fig. 8)

Installation of Ballast: It is sometimes necessary to add permanent ballast to the aircraft in order to arrive at a proper CG location. This often is the result of changing to a heavier engine or propeller, or removing or adding heavy equipment. The ballast is added to effect proper balance, not to correct for nose-up or nose-down tendencies of the aircraft. Ballast so installed must be stamped with the weight of the ballast, arm, location and date of ACA-337 Weight and Balance form covering the installation.

ACA-337: The number identifies a form supplied by the FAA for the purpose of recording repairs, alterations and weight and balance information and is a permanent part of the aircraft's papers and records. This form is for type certificated aircraft only and is not for amateur-built aircraft.

Now that you have learned the definitions of the terms used in weight and balance problems, I will explain the methods used to arrive at the desired information. It is recommended that the services of an A&P mechanic be engaged to assure that weighing and measuring is done properly, but by exercising care and checking all readings, measurements and figures as outlined in this article, you should be able to determine the CG location of your aircraft and what changes if any are necessary to bring the CG within range, providing the CG location is indicated on the plans from which you have built it. In case the CG location is not indicated on the plans, it can be determined by lengthy computations based on weight-of-material tables.

The first step in weight and balance computations is to determine the location of the datum. The most convenient location is the wing leading edge, as this point is the easiest from which to drop a plumb line for measuring purposes. A point to bear in mind when using the wing leading edge is whether or not the wing has sweepback. If it has swept-back, a point should be chosen as near the fuselage as possible.

After establishing the datum, prepare your weight and balance figure sheet as shown in Figure 7. Weighing should be done inside a closed hangar or other suitable area, free from drafts. The most suitable scales are the farm type of 1,000 lbs. capacity, or any heavy duty beam scale. Bathroom or household type scales are not satisfactory unless you use two or each side. One scale or set of scales are needed for each weighing point. The following steps are important in weighing aircraft and care should be exercised to insure accuracy.

First — check the scales on the shop floor for levelness and zero reading, then place one under each wheel of the aircraft, with the aircraft blocked up in level flight attitude. Do not set the parking brake, just chock the wheels to keep the aircraft from rolling off the scales. The aircraft must be level both spanwise and lengthwise.

Second — determine that the fuel tank(s) is empty, and

whether the oil supply is full or empty, and so indicate on the weight and balance sheet.

Third — if equipped with a water-cooled engine, all radiators and water supply tanks must be full.

Fourth — all equipment and special items such as tie-down kits, fire extinguishers, tools, etc., which are carried about as standard equipment, must be in their designated locations.

Fifth — on completion of the above four steps, recheck to determine that everything is in order and note the scale readings in the designated spaces on the weight and balance sheet. (See Fig. 7)

Sixth — remove the aircraft from the scales, without disturbing the chocks or blocks used to hold it level.

Left — Acro Sport I built by Dave Marsino, Closter, NJ. This fully aerobatic aircraft designed by Paul Poberezny can be powered with engines from 100 thru 200 hp. Can be built as a Super Acro with symmetrical wings — plans for which are available — for unlimited inside/outside maneuvers. Span 19 ft. 7 in.

Above — An unusual amphibian with engine mounted on the vertical fin, the Seawind is a 4-place composite with a cruise speed of 157 mph. Stall speed is 57 mph. Airplane is designed to accept engines up to 300 hp. Gross weight is 3200 lbs., range 915 miles.

Left — Pober Pixie, another design by Paul Poberezny. Extremely simple construction and easy and economical to fly. 50 to 75 hp.

Aircraft Rigging

by Henry E. Winslow

The Inspector has come and gone. The fabric is on hand and a brand new paint job makes the plane look ready to leap into the sky. This is it! The ship is all ready for that first test flight, almost!

Patience, lad, there is one more important detail to take care of before the first flight. Whether the plane is a brand new design or a copy of a popular project, the final adjustment of the rigging of the ship is needed. Some of the rigging, has already been built into the plane, the rest needs to be done now to insure the maximum appreciation of what the capabilities of the new plane are to be on its maiden flight.

It might be well to review a little aerodynamics so we can more fully realize just what we want to accomplish.

A flying aircraft, like a tight rope performer, is delicately balanced. The balance of the plane is around a point buried inside the ship and referred to as the center of gravity of the aircraft. Controls are attached to the plane to adjust the plane's motion around this center of gravity. The conventional plane rotates on three axis passing through the center of gravity. It **rolls** around the longitudinal axis, **pitches** around the lateral axis, and **yaws** around the vertical axis. The controls function to balance the plane on these axis as the engine propeller thrusts the plane forward.

The rudder-fin combination and the stabilizer-elevator indirectly control the attitude of the nose by moving the tail while the ailerons directly move the wings to roll the ship.

These are, together with the throttle, the primary flight controls. Airplanes also incorporate second stage controls to ease the work of flying the aircraft.

Servo surfaces and trim tabs are used to reduce the pressure needed to move the primary controls. The main need for servos has passed since the development of hydraulic boost to the controls has been adopted, but they might be remembered if a heavier ship is planned by the builder. Trim tabs are still with us, however, to ease the control pressures in long cross-country flights and in maneuvers such as turns and glides to landing or to re-rig the plane in flight to counteract the effect of CG change as fuel is used.

A third method of second stage control is being investigated. This is research into the control of the direction of the ship by shifting weights. While this has at present more application to space vehicles than to conventional aircraft it will be remembered that in the first hang gliders the body was moved fore and aft to control the rate of descent and this method was also used in the control of some early dirigibles. Even in the planes of not so long ago uncontrollable spins have been stopped, when the pilot (definitely not in the tradition of the naval captain) parted company with the spinning plane to do his own vertical descending via his parachute, when the change due to his weight brought the ship out of the spin.

The goal of correctly rigging the plane is to have it fly "hands off" in level flight at the ship's cruising speed while having a minimum of unbalance at other speeds and attitudes.

The first rigging done to the plane is seldom thought of as a step in rigging. Still the weighing of the ship to find the center of gravity and the shifting of components to bring the wings' center of pressure in correct relationship to the CG is just that.

With weight and balance completed the next step is to correct for the unbalancing forces acting on the plane in flight.

The first of these is the effect of engine torque. Another force, (and this should be mentioned together with the torque due to the similar effect), is the corkscrewing effect of the propeller slip stream. Third is to adjust the built-in stabilizing forces that correct the unbalancing effects of air currents.

In the United States the conventional engine turns the propeller counter-clockwise as viewed from the front of the engine. In a tractor configuration this causes the torque from the engine to tend to force the left wing down. To counteract this force the left wing has more angle of incidence in it than the right wing. Sometimes this is already built in as in the case of a cantilever wing. Otherwise the rear wing struts are adjustable so that the wing may be warped to counteract the torque forces. This correction solves the torque problem but creates additional drag which, in this case, turns the plane to the left. This drag is corrected by offsetting the fin to the left to balance out the additional left wing drag.

The propeller has more than once been likened to a screw moving forward in wood. This is an apt illustration because the propeller slipstream does corkscrew back and this creates an unbalancing pressure along the rear of the fuselage, as a relatively positive pressure is built up along with the left hand side of the aft fuselage and tail with a corresponding negative pressure along the right side. The correction for this is to move the fin a little further to the left.

The purpose of dihedral, sweep back, wing slots and offset engine thrust is to return the aircraft to straight and

Correct rigging can make the difference between "hands-off" stability or a tiger requiring constant control inputs that are fatiguing even on short flights. Pober Pixie shown here fully assembled for final rigging check before covering.

level flight after an air current has disturbed it. This is part of the original design of the aircraft and is a compromise between stability and controllability; i.e., the more dihedral the more control pressure is needed to bank or roll the plane. It depends, therefore, on what the builder desires in the design as to how much or how little stability is built into the ship. It is to be remembered that it is quite possible to make a ship so stable that it can be dangerous to fly. An example is the elevator trim on the Aircoupe. In this plane it is possible to, because of the limited elevator travel, trim it in a glide so that there is no more elevator control to flare the ship out with when the aircraft is ready to set down and it flies right on into the runway. Also in this ship the lack of separate rudder and aileron controls creates a hazard during crosswind landings. The more automatic controlled stability the ship has, the less control over it the pilot has. At the other end of the scale is the plane that must be flown at all times correctly!

Dihedral aids the plane to be longitudinally stable while sweep back helps to keep the plane directionally stable. The control of pitch in relation to the center of gravity of the aircraft is aided by the setting of the stabilizer at a negative angle. This causes a down load on the tail that increases as the speed of the plane increases but conversely as the speed decreases causes the nose to drop keeping the speed up above a safe limit. Often in conjunction with the negative stabilizer the engine thrust line is set down a few degrees so the down pull of the stabilizer is balanced by the down thrust of the engines. Again at slower speeds the desirable nose heavy condition results.

A higher angle of incidence in the top wing of a biplane also will control the plane's pitching at low speeds or high angles of attack as the top wing will stall first, thereby dropping the nose.

With these thoughts in mind let us go back to our own aircraft and get it ready for the first flight. All trim tabs and fixed tabs are set in neutral or slightly bent to compensate for engine torque and its effects. The wings are checked for proper angle of incidence, dihedral, and sweepback. The tail, especially the stabilizer, is checked for proper incidence (plus or minus) and the fin for offset.

Next the controls are rechecked for proper throws and direction and all stops locked. Now we are ready for the first flight.

The first part of any maiden flight is used to find out such vital facts as how well the plane handles at different speeds up to and including the stall. Also the amount of control pressures needed for fast and slow flight maneuvers; the characteristics of stalls, power on and power off, and in glides. After this part of the flight is over and if all is acceptable we are ready to check out the trim of the ship and upon landing be ready to adjust for "hands off" flying. Assuming that the flight characteristics are normal, we go to final trim. As an example let us assume that these conditions were noticed during cruising level flight: 1) Right wing tends to go down; 2) Nose goes to the left; 3) Forward pressure is needed on the stick in order to keep nose down. Upon landing and parking the aircraft we get out the tool kit and start to work. For condition 1 we will need to wash out the left wing, with the amount of adjustment depending on the amount of left stick pressure used to keep the wings level. If there is no strut adjustment then bend down the left aileron fixed tab slightly. 2) If the wash-in is decreased in the left wing do not touch the rudder until after the next flight as the change in left wing decreases the drag so the nose left tendency will be reduced. If, however, the adjustment was made by aileron trim tab, then bend the rudder trim tab to the left. To correct for 3 raise the leading edge of the stabilizer, or if it is fixed bend the trim tab on the elevator down. Another flight will give results and these adjustments are continued until the plane flies "hands off" at cruising speed.

Remember that all tabs, both fixed and movable must be positioned to the opposite direction that control is desired to be moved. That is, if the elevator is to be raised the elevator tab must be bent down. Also as each adjustment is made it is possible that the correction will set up an unbalancing action on another axis. Generally it is better to wash out a wing slightly instead of washing it in as in this way the aileron control in a stall is improved.

On the first flight it is very important to check the action of the plane in a stall, especially if the ship is a low or midwing design. If the stall is quite violent with a lot of tail shake, a check on the wing fuselage fairing should be made to check for blanketing of the tail surfaces by the wing in the stall. Often just a small change to redirect the down wash from the wing will be all that is needed. In any case, no aerobatics should be attempted until a small, easy stall has been rigged into the plane.

A word about fairings in general. The more streamlining that is done to the plane, the better the performance will be. A trip to a used plane boneyard often uncovers adaptable fairings for your ship. If none are to be had, an interesting project is the building of molds and the fabrication of fiberglass fairings. However, that is the subject for an article in itself.

As the plane is being built it is wise to make up some rigging tools. For any plane a wing chord incident bar is almost mandatory as the angle of incidence is checked again and again. Control throw quadrants are well worthwhile also and they allow one person to set the control stops. By clamping the quadrant to the fixed surface wing stabilizer or fin and by using a smalll size shock cord or even a length of inner tube as a bungee on the control, the stop may be set to the proper throw. One of the handiest times to use on aileron or elevator is the inexpensive "devil's level". Due to its construction it will not be accurate over 30 degrees from level but within this range if any eye loupe is used one can read to ¼ degrees very accurately.

It is also suggested that tables be made up giving the angle of incidence of the wing, control throws, flap travel, gap and stagger in the case of biplanes, etc. Also the datum point used for weight and balance figuring, where the leveling lugs are located, what degree of dihedral the wing has. In fact, a small notebook should be used to make up an aircraft manual. The tables mentioned above should be included plus the center of gravity travel, type of wing chord used, the operations limitations as to airspeed and engine. In fact, it would be well to have all pertinent construction and operational data included. This manual should be kept in the ship and if the plane is sold it could mean a great deal of time saved by the new owner in overhauling the ship. In fact, even though the plane is kept by the builder, time has a way of allowing many of these details to be forgotten and if no record is available a lot of work is involved in remeasuring, especially if those parts are buried behind the fabric of wing and fuselage. The **Amateur-Built Aircraft Service & Maintenance Manual** will help. It is a manual tht should be kept in the airplane.

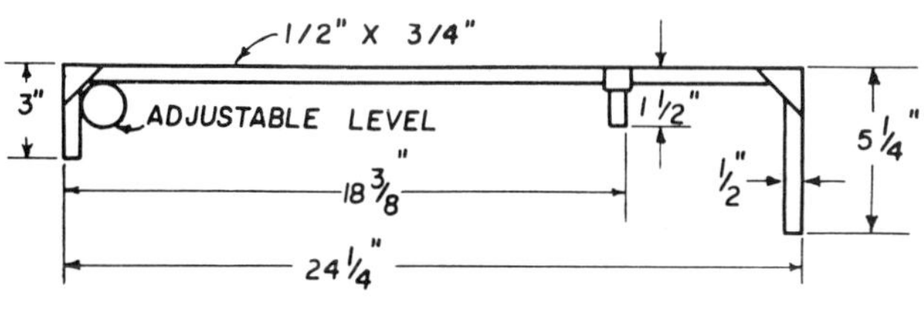

A simple tool, easy to construct, for checking wing incidence along wings (wash-in and wash-out). Dimensions shown for either Pitts or Acro Sport I — modify specs as necessary for other aircraft wings.

Powerplant Installation

by David B. Thurston

Powerplant installation warrants the attention of the amateur builder and designer since problems in this area are responsible for 25% of light aircraft accidents. Part of this discussion is intended to show how the powerplant and fuel system can be made more foolproof; thereby placing fewer demands on the pilot and making the aircraft as safe as possible. Although my experience has been mainly with type certificated engines and many references are made to them, the same safety suggestions apply to automotive engines converted for use in homebuilt aircraft. There are some obvious advantages in using a certificated engine such as dual ignition and the availability of known cooling limits and cooling standards for design purposes. For instance, Lycoming and Continental tell you the number of cubic feet of air per minute that must flow over the cylinders as well as required oil, magneto, and alternator temperature limits.

First, make sure you have ready access to various important parts of your installation such as oil drains and screens, accessories, mags, carburetors, etc. This may sound obvious, but is not always easily achieved. For example, if the oil screen comes out only as far as an engine mount cross member and no further, then you probably won't be able to service the screen. The result is a clogged screen and engine trouble.

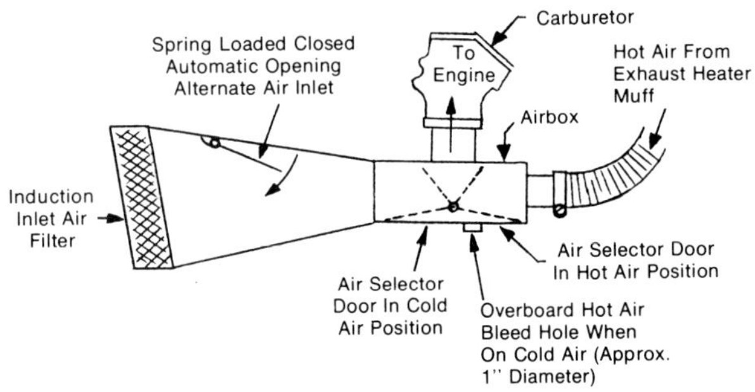

FIGURE 1. BASIC INDUCTION SYSTEM (CARBURETOR ENGINE)

It is important to use dynafocal mounts on the engine if possible to keep engine vibration isolated from the rest of the airframe structure. This reduces fatigue on the structure and, just as importantly, reduces cabin fatigue. A pilot who is less tired is less apt to make poor decisions. When building the 1908 June Bug replica, we actually used dynafocal mounts to help reduce torque loads on the very light airframe. Incidentally, for a good guide on mounting engines and other pertinent facts write to: Lord Kinematics Div., Lord Corporation, Erie, Pennsylvania 16512 and ask for Technical Reference No. LB-571 "General Aviation Engine Suspensions".

As another safety feature, place fuel in the wings rather than the fuselage. This location reduces the fire hazard in case of accident and also reduces the loads on wings in flight. In small planes, you might use two tanks and interconnect them with a 3/8" diameter vent and same size cross-feed so they act as a single tank. You should provide a selector valve which will shut-off either tank individually in case of a ruptured tank in flight; requiring a selector with LEFT, RIGHT, BOTH, AND OFF positions. Make sure that operation of the fuel selector arm has no play and be sure you can easily see where the selector is pointing. Also have positive detents to firmly hold the valve in the selected position.

A number 16 mesh stainless steel or brass finger strainer protruding into the fuel tank should be used to prevent system clogging from foreign matter that may get into the fuel tank. The strainer should be 5 times as long as its diameter. Check and clean the strainer every 100 hours or six months — whichever comes first. Do not paint the inside of your fuel tank since paint will peel and clog the fuel system.

Although not required by the FAA, some accidents might be averted if a spring-loaded door were provided in the induction system to act as an automatic alternate air source in case the airscoop or filter suddenly becomes clogged by ice or debris such as paper. A spring-loaded door which requires about 0.5 pound applied at the center to open should be used so alternate air can automatically enter the induction system if necessary. This arrangement is shown in Figure 1.

Hot air should not become stagnant in the heat muff when carburetor heat is shut off. To provide flow, about a one inch diameter bleed hole should be located somewhere in the air box to allow hot air to flow continuously and not burn out the system. This hole can act as a drain for the air box so excess fuel from the carburetor cannot become trapped and catch fire from possible back-firing during starting. Arrange this hole so it is closed when carb heat is turned on as shown in Figure 1. The FAA requires that the hot air system provide at least 90°F temperature rise in the carburetor throat to insure the removal of ice. Test your system with a temperature probe located at the outlet of the air box to measure carburetor inlet temperature at max climb power and at cruise. Run this test in approximately 30°F, waterfree air conditions.

The cylinder head temperature gauge is another safety feature that deserves consideration. Usually a gauge on one rear cylinder is enough, such as the left rear (#4 on Lycoming engines). Also, leave a couple inches of vertical baffle clearance behind the rearmost cylinder so the airflow can use this volume as a plenum chamber extending down around the cylinder. Number 4 will probably be your hottest cylinder. If you bring the baffle too close so there is no plenum, the back side of #4 will be starved for air. Dropping the vertical baffle back 2" will permit better cooling of that cylinder.

Use steel fittings and steel or fireproof fuel and oil lines inside the engine compartment. They will last longer in the event of a powerplant fire.

FUEL FLOW GROUND TESTS

To make sure adequate fuel is delivered to the engine, some rather simple tests should be run before a new airplane is flown. Fuel flow is affected by a number of factors such as fuel line size, fittings, number of bends, fuel tank location, and venting.

Lycoming engines require that fuel pressure be at least 0.5 psi at the carburetor. Most aircraft engines are 2.5 to 4 psi. To check the flow in a gravity flow system, remove the fuel line from the carburetor and raise the open end 20 inches (since 20 inches height of fuel give 0.5 psi). The calculated fuel flow noted below should be delivered with the fuel line held in this position.

The flow rate required depends upon the horsepower of the engine, and would be about 0.6 lb. per brake horsepower

Note: Much of the material presented here is based upon Chapter 11 of Thurston's book "Design for Flying", McGraw-Hill 1978.)

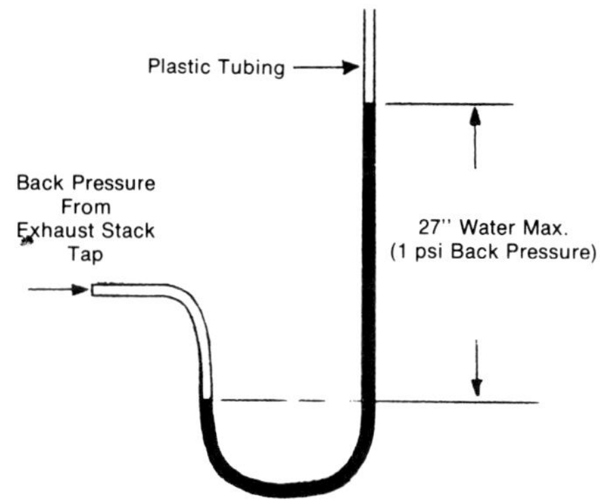

FIGURE 2. A SIMPLE MANOMETER TO MEASURE EXHAUST BACK PRESSURE

per hour. With the aircraft positioned in the maximum climb attitude, turn on the fuel selector valve and allow fuel to flow into a container. A 100 hp engine would need 60 pounds or 10 gallons per hour. Ten gallons of fuel per hour is maximum take off flow for an 0-200, usual engineering practice is to multiply that times 1.5 to make sure you have proper flow which is 15 gallons per hour or one gallon for 4 minutes per 100 hp. Time the flow for 2 or 3 minutes, running this test several times for better accuracy and to average the measured flow. If you have an engine driven pump which requires engine operation while doing this test, use an electric auxiliary pump to supply fuel to the engine. Disconnect the engine pump on the carburetor side of the pump at the carburetor. Again lift the fuel line up 20" and measure flow rate from the engine driven pump. If you have both an engine driven pump and an electrically driven auxiliary pump you should run separate tests for each system.

Another important test is the determination of exhaust back pressure. If there is too much back pressure at the exhaust valve ports, the valves could burn away. To prevent this, the exhaust back pressure should not exceed two inches of mercury, which is about one psi or 27 inches of water. (Water is heavier than gasoline). On the longest exhaust stack drill a #40 hole just below the flange, weld a boss around the hole, and tap for a fitting. Connect a line from this tap to a simple water manometer which can be nothing but an open ended piece of plastic tubing taped to a yardstick in a U shape. Make sure the differential between the two water levels is no more than 27 inches at full throttle as shown by Figure 2. Don't forget to document all of these tests in the engine logbook.

COOLING SYSTEM CALCULATIONS

We shall now discuss how you can calculate the size of inlets and outlets for an air cooled engine cowling. Of course, our modern air cooled engines are cooled almost as much by fuel and oil as by air flowing over the cylinders, so air flow through the oil cooler must also be taken into consideration.

The pressure, velocity and temperature of the air at the inlet are designated P_1, V_1, T_1, while outlet air parameters are designated P_2, V_2, and T_2. As air flows through the engine it picks up heat, expands, and comes out at a velocity that is preferably slightly greater but may be slightly less than the free airstream, velocity, V_1. For simplification, we shall ignore some of the less significant variables. Solving for the relationship between inlet and outlet airflow, we can determine how much larger the outlet area should be than the inlet to provide an optimum flow through the cooling system. The reason for doing this being that too large an outlet will pull too much air through the system and thus cause an unnecessary amount of drag; sacrificing performance in the process. Too little inlet area of course will cause engine overheating. First we'll calculate inlet area.

Engine manufacturers specify the amount of cooling airflow through the engine in cubic feet per minute. Assume, for example, a 2700 ft.^{3} cooling air requirement, a 20 in.^{2} oil radiator face, and a climb speed of 70 mph. (You should use the optimum climb speed for your aircraft). Other conditions are a standard air temperature of 59°F and pressure of 29.92" mercury.

Converting to ft./sec. $\dfrac{2700 \text{ ft.}^3\text{/min.}}{60 \text{ sec./min.}} = 45 \text{ ft.}^3\text{/sec.}$

Converting 70 mph to ft./sec.:

$70 \dfrac{\text{mi.}}{\text{hr.}} \times \dfrac{88}{60} = 102.7$ ft./sec. (where $\dfrac{88}{60}$ converts mi./hr. to ft./sec.)

Assume losses of 30% at the cowling inlet, giving an entrance flow coefficient of 0.7. Now solve for inlet area:

$$\text{area} = \frac{45 \text{ ft.}^3\text{/sec.}}{102.7 \text{ ft./sec.} \times 0.7 \text{ coef}} = .626 \text{ ft.}^2 \text{ for cooling cylinders}$$

$$\text{oil cooler} = \frac{20 \text{ in.}^2}{144 \times 0.7} = .198 \text{ ft.}^2 \text{ in required inlet area.}$$

This gives a combined inlet area of .824 ft.^{2} on a standard day. But the critical condition will be a 100°F day. To correct .824 ft.^{2} for a 100°F day we must multiply by the density ratio for 100°F temperature: Standard density = .002378; 100°F density = .00221 (from air density charts at sea level). Therefore:

$$\text{Area at } 100°\text{F} = .824 \times \frac{.002378}{.00221} = .887 \text{ ft.}^2\text{, or } .9 \text{ ft.}^2.$$

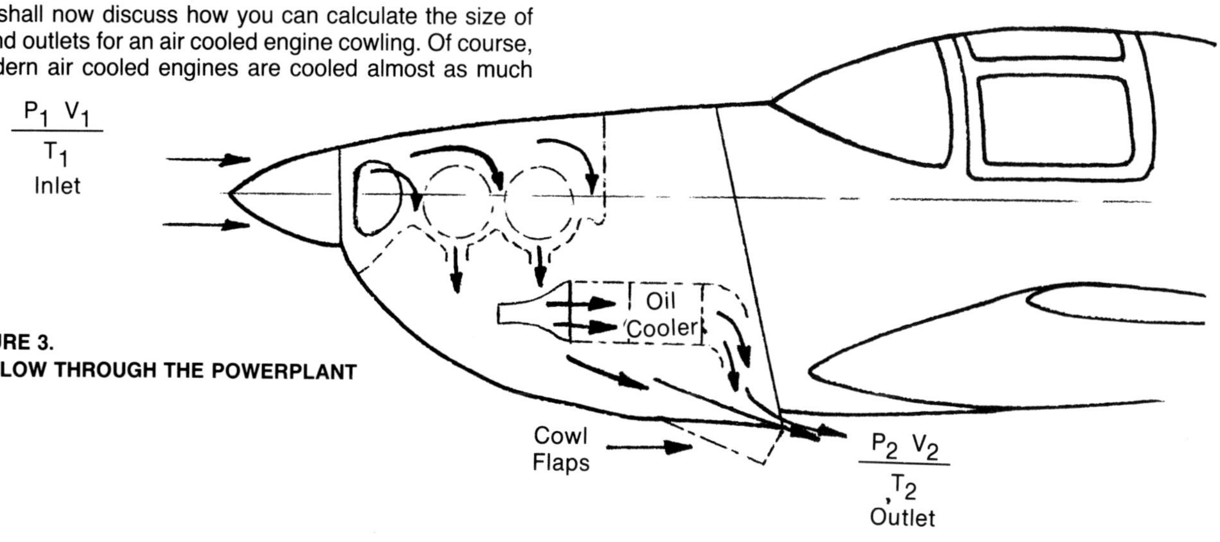

FIGURE 3.
AIRFLOW THROUGH THE POWERPLANT

Thus, .9 square feet should be adequate inlet area for a 100°F day climb at sea level. If you have two openings in the cowling, one on either side of the propeller, the total of the two openings should be .9 sq. ft. If you have a single opening, there will be some dead airflow around the spinner or prop hub on either side. The .7 coefficient takes care of the dead loss around the entrance to the edges of the opening but there is another loss associated with disturbed flow over the hub area. Consider this disturbed area to be 1" thick and measure the required area starting 1" outboard from the spinner or prop hub.

OUTLET

The cooling air exit area must handle the inlet airflow plus the expanded volume caused by removed engine heat. If we use the 100°F day with standard 14.6 psi inlet pressure, then $P_1 = 14.6$. The cooling air can be expected to pick up about 100°F temperature rise when passing over the cylinders. This temperature must be in absolute units which is ambient temp plus 460°F. T_1 = 100°F day + 460°F = 560°F. It is common to assume outlet pressure of 14.1 psi giving .5 psi differential. T_2 absolute temperature will be 100°F + 460° + 100° rise over engine = 660°F.

$$P_1 = 14.6 \qquad P_2 = 14.1$$
$$V_1 = 1 \qquad V_2 = ?$$
$$T_1 = 560°F \qquad T_2 = 660°F$$

$$\frac{P_1 V_1}{T_1} = \frac{P_2 V_2}{T_2}$$; solving for V_2 in terms of V_1:

$$V_2 = V_1 \left(\frac{T_2 P_1}{T_1 P_2} \right)$$

$$V_2 = V_1 \left(\frac{600 \times 14.6}{560 \times 14.1} \right) = 1.22 V_1$$

Thus we require 22% more exit area than entrance area for this set of conditions. So if 128 sq. in. (.887 sq. ft.) is the basic entrance area, then the exit should be 1.22 × 128 sq. in. = 156 sq. in. for climb conditions.

For the cruise condition, solve the equation again using cruise velocity instead of 70 mph climb velocity to establish inlet and outlet areas. For a high performance airplane much less exit area will be required. The inlet area cannot be easily reduced since it is determined by climb requirements. However, the minimum exit area calculated for cruising flight should be used to obtain optimum cruising speed performance. In order to optimize exit area cowl flaps may be required to provide enough exit area for climb flight. In which case design the exit area for the cruise condition, opening the cowl flaps during climb. On a cold day it probably will not be necessary to operate the cowl flaps.

This brief summary should be of assistance in making your powerplant installation a bit safer, while the use of proper cooling airflow will increase engine life as well as offer optimum cruising performance and fuel economy. And so I hope this information will prove useful in developing safer and more efficient homebuilt aircraft.

Approximate Engine Weights

Continental

Type	Weight
A50, 65, 75, 80(-8)	170-177 lbs.
C85-8, C90-8	158-159 lbs.
C75-12, C85-12	168 lbs., approx.
C90-16F, 0-200	170 lbs. without electrical 212 lbs. with electrical

Lycoming

Type	Weight
0-235	240-255 lbs.
0-290	260-265 lbs.
0-320	268-285 lbs.
IO-320	280-294 lbs.
AIO-320	306-307 lbs.
0-360	282-301 lbs.
IO-360	294-332 lbs.

Volkswagen

Type	Weight
Minimum Engine	117-145 lbs.
Monnett Conversions	151 lbs.
HAPI 1600cc and 1834cc	152 lbs.
Revmaster, 2100cc, Roller Crank	178 lbs.
Revmester, 2100cc, Babbit Mains	160 lbs.
Limbach 1800cc	164 lbs.
Limbach 2000-2600cc	201-205 lbs.

The above weights include carburetors, magnetos, ignition harness, spark plugs, tachometer drive, starter and generator/alternator drives and starter and generator or alternator. The only exception is the Volkswagen engine listed as the "minimum engine".

Engine and propeller attaching parts will run another 3-5 pounds and if you remove the starter and generator or alternator this will remove about 21 to 43 pounds in most aircraft style engines. It is suggested that if you do take these off that you weigh them. Also, for electrical system accessories add about another three pounds for an engine driven fuel pump at about 2-3 pounds. It is recommended that engines used in aerobatics have the solid crank shaft. Don't forget to weigh the engine baffles! You do need to add the inverted oil system weights of components lines, fire shields, etc. It is recommended for accuracy that you put your engine on the scale with all the accessories from the fire wall forward that will be used. It is also recommended that you do this before you build a mount. The Acro Sport II engine, (0-360), on N9AS weighs about 288 lbs. without the starter and generator. Lastly, if you are building an engine from matching parts, get the service bulletins or service bulletins, letters and instructions from the engine manufacturer for aircraft engines. The use of a good Volkswagen book is recommended for those building thir own Volkswagen engine.

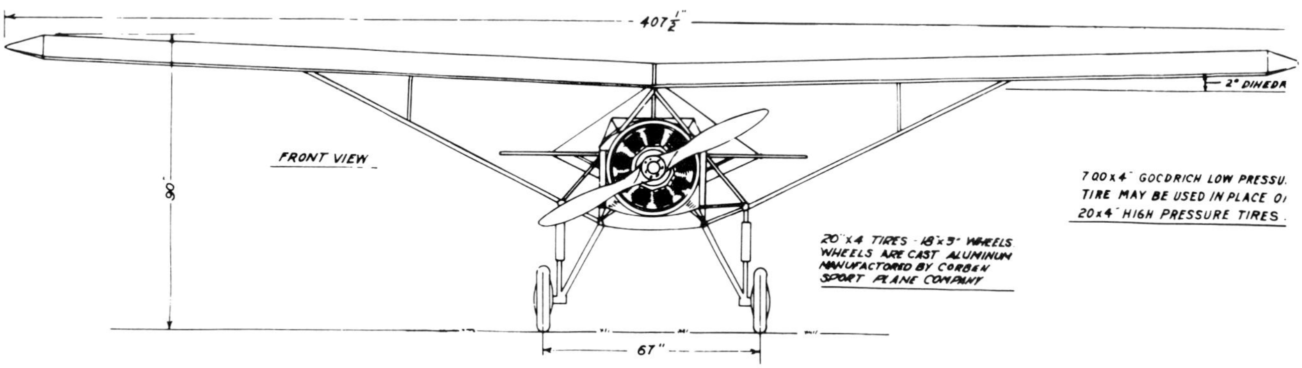

Corben Junior Ace — A Homebuilt Of The 30's

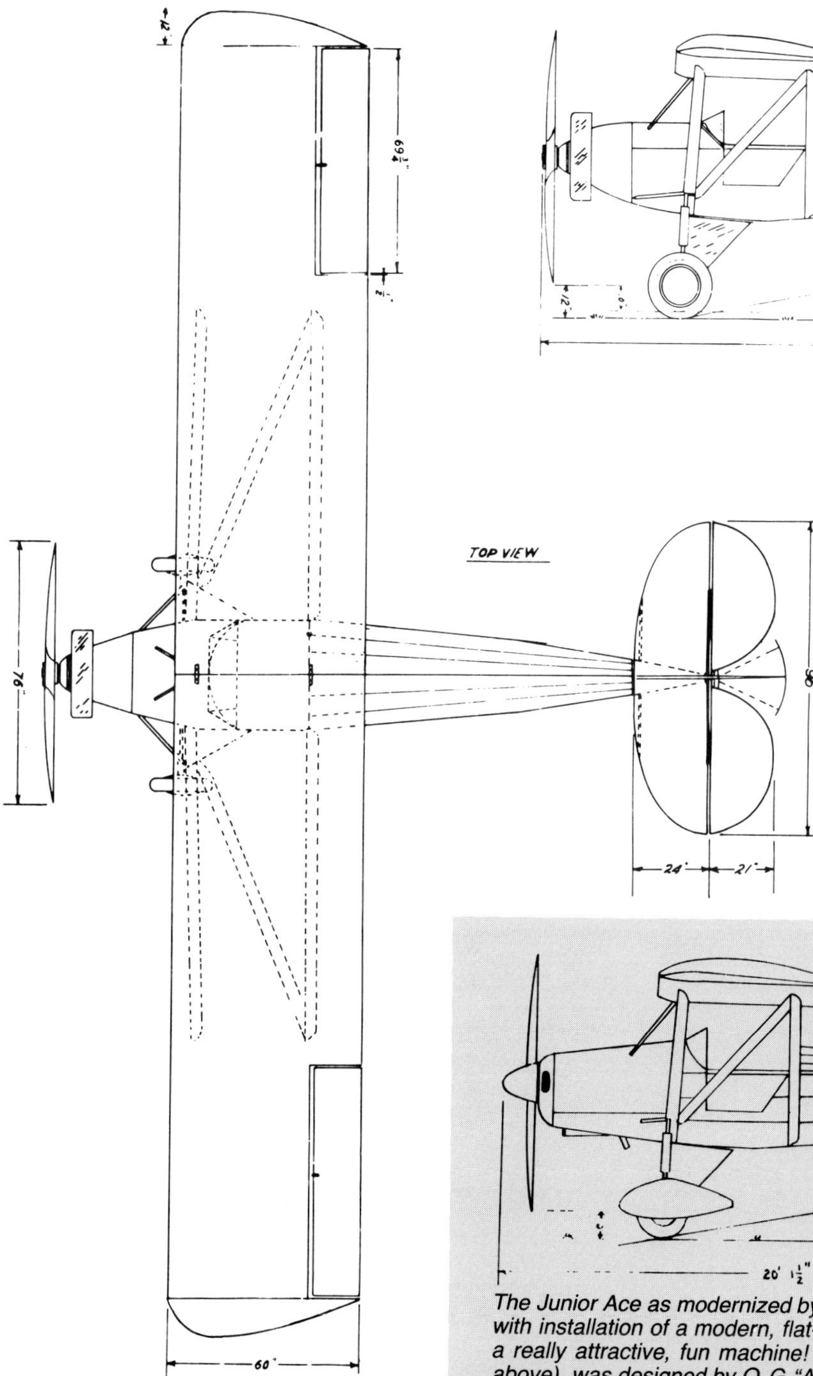

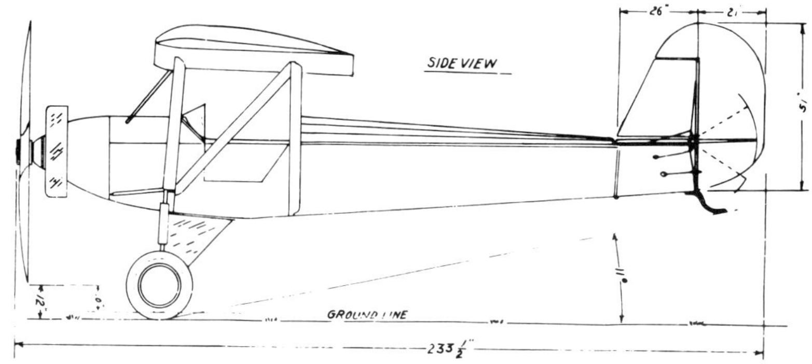

Wing Span	407½ in.
Height	90 in.
Length	225½ in.
Weight Empty	563 lbs.
Useful Load	412 lbs.
Gas Capacity	10 gals.
Oil Capacity	6 qts.
Total Wing Area	168 sq. ft.
Total Aileron Area	19.42 sq. ft.
Total Stabilizing Area	13.82 sq. ft.
Total Elevator Area	10.03 sq. ft.
Total Fin Area	3.34 sq. ft.
Total Rudder Area	5.18 sq. ft.
Airfoil Clark-Y	Chord 60"
Angle of Incidence	2 degrees
Dihedral	2 degrees
Rudder Travel	50 degrees
Elevator Travel	60 degrees
Wing Loading	5.80 lbs./sq. ft.
Power Loading	19.5 lbs./h.p.
Power Plant Aeromarine A-R-3	50 h.p.

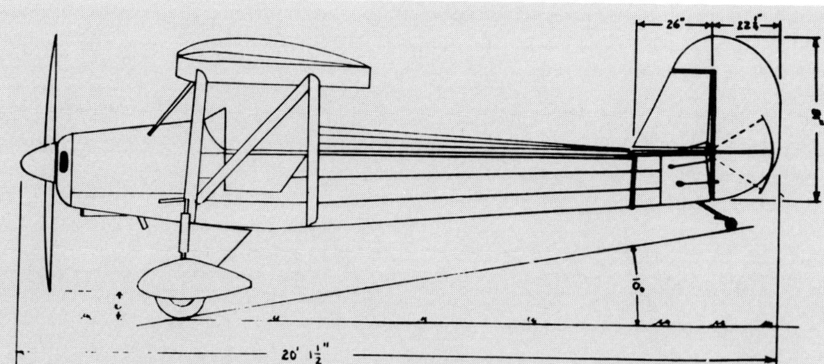

The Junior Ace as modernized by Paul Poberezny. More attractive lines are evident with installation of a modern, flat-four engine and more efficient cowling. Makes for a really attractive, fun machine! The original Junior Ace (as shown in 3-views — above), was designed by O. G. "Ace" Corben in the early 30s. Corben also designed the Baby Ace and Super Ace — favorites of today's homebuilders. Plans for the Junior Ace were published in installments in the 1933 issues of Popular Aviation (Flying).

Installing Wood Propellers

by Robert Bristol

TABLE NO. 1
OPTIMUM WOOD PROPELLER INSTALLATION HUB COMPRESSION METHOD

Hub Thickness(in.)	Desired Hub Compression	Torque Bolts	(Bolt Rev.)
$3\frac{3}{8}$	0.020	$\frac{3}{8}$-24	0.49
		$\frac{1}{2}$-20	0.41
4	0.024	$\frac{3}{8}$-24	0.58
		$\frac{1}{2}$-20	0.48
4	0.029	$\frac{3}{8}$-24	0.68
		$\frac{1}{2}$-20	0.57
$5\frac{3}{8}$	0.032	$\frac{3}{8}$-24	0.77
		$\frac{1}{2}$-20	0.65

TABLE NO. 2
WOOD PROPELLER INSTALLATION TORQUE WRENCH METHOD

Size of Steel Specification	Aircraft Bolts Dia. (inches)	Recommended Wrench Torque (in.-lb.)
AN6	3/8	200 ± 25
AN7	7/16	250 ± 25
AN8	1/2	300 ± 25

Ref. AN 01-1A-13(1946)

Caution: Over-tightening propeller attaching bolts will cause the wood of the hub to crush, breaking its moisture seal and slightly reducing drive-torque capacity of the installation.

TABLE NO. 3
ADAPTER RETENTION BOLTS RECOMMENDED WRENCH TORQUE

Size of Steel Specification	Aircraft Bolts Dia. (inches)	Recommended Wrench Torque (in.-lb.)
AN6	3/8	280 to 300
AN7	7/16	480 to 540
AN8	1/2	720 to 780

Installation of a wood propeller will require a front face plate of adequate stiffness (approximately the same size as the propeller hub), a flange adapter (in some cases), and a set of attaching bolts of the proper length.

It can be shown that an engine must deliver its driving torque to a wood propeller through static friction. That is, the force which resists movement of the propeller hub on the engine flange is due to compression of the wood surface against the flange. Therefore, it is important to compress the wood to its maximum during propeller installation, but also important to avoid crushing the wood. Although the drive bushings incorporated in most flanges provide a back-up system, a load will be imposed on them only if there is movement of the propeller on the flange. The bushings can carry engine driving torque loads for only a short period of time.

The data from Forest Products Laboratory, U.S. Department of Agriculture, Madison, Wisconsin for Yellow Birch wood shows that the optimum compression pre-load of a wood propeller hub is approximately .006 inch per inch of hub thickness (i.e., a propeller hub which is one inch thick before installation should be compressed .006 inch to a thickness of .994 by drawing up the propeller attaching bolts). Knowing the hub thickness and the number of theads in each inch of attaching bolt thread, it is possible to calculate the proper additional rotation of each attaching bolt after the front face plate, propeller and flange (or adapter) are snug. For example, a propeller hub which is 4.50 inches thick should be compressed .006 × 4.50 = 0.027 inch. If ½ inch dia. bolts are used (½ × 20 UNF threads), they should be turned 0.027 × 20 − .54, or just over ½ revolution after the front face plate, propeller and flange are snugged. See Table No. 1 for an example of common installations.

Generally recommended wrench torque to achieve proper hub compression against standard flanges are listed in Table No. 2. These torque recommendations do not consider variations of thread condition and assume the threads of the bolts and drive bushings are clean and dry.

INSTALLATION PROCEDURE

After study of the preceding discussion of wood propeller installation requirements of your propeller assembly, the following installation procedure should be followed:

(1) Be certain that the magneto switch is "off" and both magnetos are grounded.

(2) Install the flange adapter, if required. Adapter retention bolt wrench recommendations can be found in Table No. 3. Install safety wire in heads of adapter retention bolts in pairs, following the safety wire groove provided in the adapter.

(3) Locate the propeller on the engine flange or adapter, in the most convenient position for hand cranking.

(4) Remove a spark plug from each cylinder. Chock the aircraft wheels to prevent movement.

(5) Install the propeller attaching bolts "finger-tight" so that the faceplate, propeller and flange are snug (but the wood has not been compressed). Check track of the blade tips by rotating the tips past some fixed object on the floor. The tips must track within 1/16 inch of each other when the installation is complete. Track should be corrected at this time by snugging up the bolts nearest the blade which is complete. Track should be corrected at this time by snugging up the bolts nearest the blade which is forward. This will result in a common starting point for all the attaching bolts.

(6) Tighten the attaching bolts in small increments, moving diagonally across the bolt circle. It is good practice to check track frequently while tightening the bolts. Take care to tighten bolts on opposite sides of the blade centerline evenly so that blade-to-blade conformity of angles is maintained.

(7) Since a small part of the wood compression is plastic (permanent) it is a good practice to loosen the bolts to allow the wood to relax for an hour. Re-tighten following the same procedure.

(8) Install safety wire. It is a good practice to wire the attaching bolts in pairs (not a continuous wire) twisting the wire between bolt heads.

OPERATING TIPS

The following practices will add to the service life of your wood propeller.

(1) Do not use the propeller as a tow bar to move your aircraft.

(2) Avoid running up in areas containing loose stones and gravel.

(3) Place the propeller in a horizontal position when parked.

(4) Inspect frequently for bruises, scars, or other damage to wood and blade leading-edge protection. It is a good practice to conduct pre-flight and post-flight inspections.

(5) Protect your propeller from moisture by waxing with an automotive type paste wax. Keep the drainholes in metal tipping open.
(6) Assume that your propeller is unairworthy after any kind of impact until it has been inspected by qualified personnel.
(7) Inspect and check propeller attaching bolts for tightness at least every 100 hours or annually. More frequent inspection may be necessary when climatic changes are extreme.
(8) All wood and metal tipping repairs must be made at the factory or by an approved propeller repair station.
(9) Check propeller balance whenever there is evidence of roughness in operation.

If your propeller begins to show any of the following damage, it should be retired from service:
(a) Cracks in hub bore.
(b) A deep cut across the wood grain.
(c) A long, wide or deep crack parallel to the grain.
(d) A separated lamination.
(e) Oversize or elongated hub bore or bolt holes.
(f) An appreciable warp (discovered by inspection or through rough operation)
(g) An appreciable portion of wood missing.
(h) Obvious damage or wear beyond economical repair.

Refer to FAA publication AC43.13-1a for further information.

Forward cowling removed, Pixie discloses many details in its construction. Note how simple a design this is! Nothing complicated. Easy to build. Easy to maintain. And the VW engine is easy on the gas. Makes for lost cost flying!

What You Should Know About Load Factors

by Bob Whittier (EAA 1235)

Studies made by federal aviation authorities of civil aircraft airworthiness requirements and accident records indicate there is a persistent tendency among pilots to use small airplanes for acrobatic maneuvers, for which the planes were not originally designed. While airworthiness requirements in general aviation afford some protection against widespread structural failure accidents, it is still essential that designers, builders and pilots of all kinds of airplanes be well informed on the strength limitations of their aircraft. The government, therefore, prepared a report containing a simple explanation of the basis on which airplane strength requirements are determined, which shows how flight maneuvers should be limited to avoid structural failure. The text of this report follows.

How Strong Should An Airplane Be?

This question confronts every airplane designer and is one of the most difficult of all engineering problems. Because of its importance from a safety standpoint, the question is actually answered by the government, which sets up minimum strength standards for all commercial airplanes. The FAA insures that every certificated airplane meets the minimum strength requirements. Some commercial airplanes **exceed** the required strength, but in general the requirements are followed quite closely. Every pilot should be interested in knowing something about the manner in which the strength of an airplane is determined and what can be expected of the structure in flight.

Airworthiness requirements must be carefully worked out so as to result in **efficient** as well as **safe** airplanes. This is an extremely difficult technical problem as an efficient airplane must be light, while a safe airplane must be strong. Extra strength means extra weight, which means reduced payloads. An airplane, unlike a bridge or building, cannot afford to have any excess structural weight beyond that which is essential for safety.

In writing the airworthiness requirements which determine the airplane's strength, the FAA must therefore make every effort to determine, in advance, the worst loads likely to be put on the airplane during its lifetime. Here we run into a difficult problem, because the worst possible loads are much too high to use for efficient design. Any pilot can make a very hard landing in loads that might be called abnormal. For that matter, he might even fly the airplane into a brick wall. These abnormal loads must be ignored entirely if we are to build airplanes that will take off quickly, land slowly and carry a good pay load. But we still have to decide where to draw the line between normal and abnormal loads. And, having decided, we must try to see that pilots are aware of the fact that abnormal loads are not provided for in the design of most commercial airplanes. The purpose of this report is therefore to help pilots understand where the line is drawn between these so-called normal and abnormal loads, and to connect this up with actual flight maneuvers as much as possible.

How are design loads determined?

Any pilot knows that an extremely hard landing may break the landing gear. It is not so generally appreciated, however, that a hard landing may produce hidden damage in the landing gear or in members which serve to carry flight loads. Hidden damage from hard landings may result in the subsequent failure of some part carrying flight loads, even though the airplane is flown gently. (This has been repeatedly borne out by accident reports). In view of this it should be regarded as imperative that the airplane structure, particularly wing truss structure directly affected by landing loads, should be subjected to a rigid inspection after any abnormally hard landing.

With these words of caution in regard to landing loads, we can now devote our entire attention to the loads produced in flight. These are mainly applied to the wing, as it is the wing that supports the weight of the airplane against the pull of gravity. In level flight the net result of all the air pressures acting on the wing is an upward load just about equal to the entire weight of the airplane (it would be exactly equal if there were not air loads acting on fuselage or tail surfaces).

Aeronautical engineers have adopted a simple method of stating the value of the air load acting on the wing. Instead of giving you this value in pounds, they use the term **load factor.** The load factor is simply the ratio between the total air load on the wing and the weight of the airplane. Thus when the wing is producing a "lift" equal to twice the weight of the airplane, we say that the **load factor** is 2. Load factors are also used in talking about landing conditions, in which case we deal with the load on the landing gear, instead of the wing. In a hard landing the total load acting upward on the wheels may be as much as three times the weight of the airplane. The **landing load factor** in this case would be 3.

When an airplane is in flight there is, of course, no support from the landing gear so we must deal entirely with the wing in talking about load factors. As soon as the airplane has left the ground, the weight of the entire airplane must be supported by air pressure acting on the wings. (Actually this pressure is transmitted to the ground by the air so we could truthfully say that the weight of the airplane really never leaves the ground). In level flight the total "upward" air load on the wings just equals the weight of the airplane. The **load factor** is therefore one for level flight. In a vertical dive, however, the wings no longer have to lift the airplane, so the load factor is zero, or very nearly so. We will see later what sort of conditions cause load factors greater than one.

Now, if a designer tells you that his airplane is designed to 6 load factors, he means that the wings are designed to take a load equal to 6 times the weight of the airplane. Usually this means that the wing will **break** at a load factor of 6, in which case we say that 6 is the ultimate load factor. This does not mean, however, that it is safe to put this load on the airplane in flight. A structure that breaks at 6 load factors will usually begin to take permanent set or distortion, or show others signs of distress at a considerably lower load factor. The load factor at which permanent set begins to take place is called the yield load factor. This is usually about two-thirds of the ultimate load factor. In the above case where the ultimate load factor was assumed to be 6, the yield load factor would then be 4.

All commercial airplanes below 2,000 pounds gross weight are required to be designed to ultimate flight load factors of at least 6. (Some airplanes in this range run as

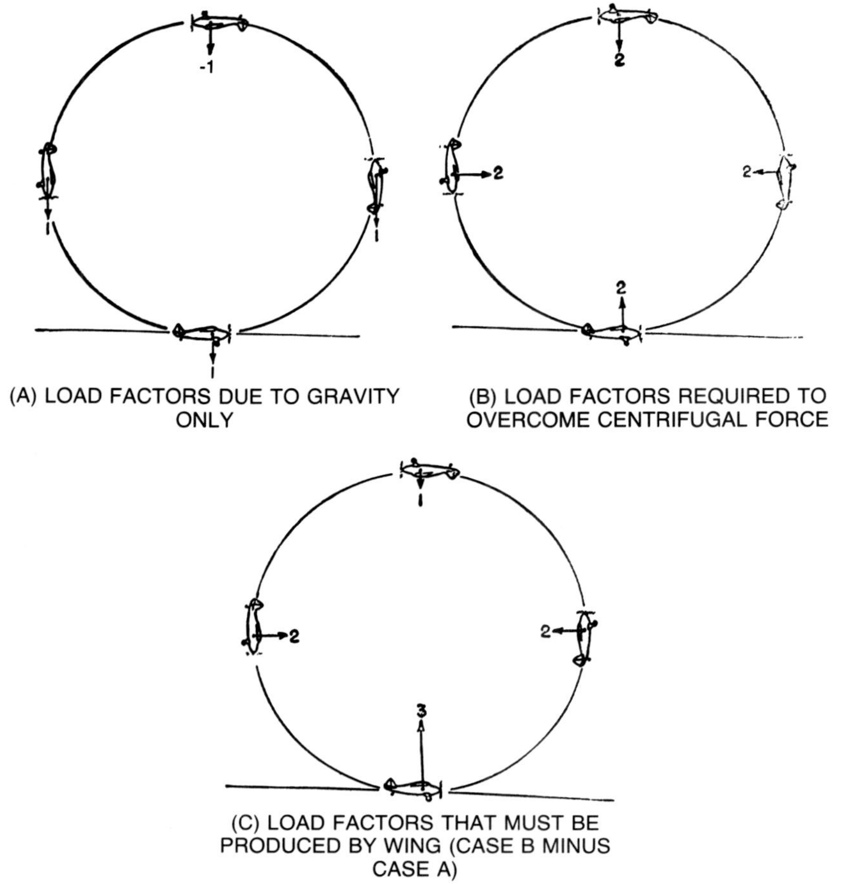

FIG. 1 Load factors in "perfect" loop at constant speed.

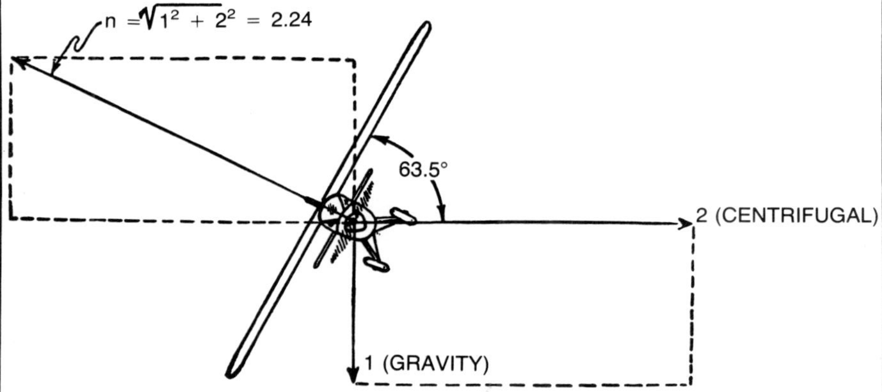

FIG. 2. Load factors in speed turn.

Angle of Wings to Horizontal	Example	Load Factor Required
0 degrees		1
10 degrees		1.01
20 degrees		1.06
30 degrees		1.15
40 degrees		1.31
50 degrees		1.56
60 degrees		2.0
70 degrees		2.92
80 degrees		5.75
90 degrees		Infinity

TABLE I — Load Factors in Steep Turns

Angle of wings to horizontal	Percent increase in normal stalling speed	Actual stalling speed based on 50 miles per hour normal stalling speed
Degrees		**Miles per hour**
0	0	50
10	.5	51
20	3.0	52
30	7.0	54
40	14.4	57
50	25.0	62
60	41.4	71
70	71.0	85
80	240.0	120
90	Infinity	Infinity

TABLE II — Minimum Speed for Banked Turns

Accelerometer

high as 7 or more.) We can therefore use 6 as a starting point in discussing load factors for small airplanes.

We have seen already that an ultimate load factor of 6 will give a yield load factor of about 4. To be perfectly safe against any yielding of the structure we should probably allow some more margin to take care of depreciation, poor workmanship, repairs, etc. On this basis the maximum recommended operating load factor would usually be somewhat less than 4. A value of 3 would be low enough to cover almost all cases and will therefore be used as a basis for discussing the operating conditions.

Physical significance of a load factor

From a strength standpoint we have seen that a load factor simply means the ratio of one load to another, the latter load being the gross weight of the airplane. Thus the recommended operating load factor of 3 represents a flight condition in which the load on the wing is three times the weight of the airplane. We can now consider how such a condition might be obtained in actual flight, and what it means to the pilot.

We know that a load factor of one is required to hold the airplane up in level flight. A load factor of 3 therefore represents an excess or unbalanced load factor of 2. What does this do to the airplane and to the pilot? As far as the airplane is concerned it acts like any other body subjected to an "unbalanced" force — it accelerates in the direction of the force. This means that it picks up speed in an upward direction. But since the airplane is already traveling in a horizontal direction, the flight path appears to curve upward. In general, then the effect of a load factor is to cause a **curved flight** path.

This is more easily seen by working in the other direction and considering what the pilot does when he makes a pull-up. While flying level, he pulls back on the stick, causing the tail to go down and the angle of the wing to increase. This increase angle of attack of the wing produces an additional lift, which causes the airplane to accelerate upward and follow a curved flight path.

At this point it might be well to mention that a load factor can also be thought of as the ratio between a given load and the pull of gravity. Since the latter is actually measured by the weight of an object (in this case the airplane), we can see that there is no discrepancy in the two definitions. In fact, it is common to speak of a load factor of 3 as 3 "g", where "g" refers to the pull of gravity. Strictly speaking, "g" refers to the acceleration of gravity and should not be used in talking about load factors. The acceleration of the airplane, as we have seen, depends on the net load factor; i.e., that part of the total load factor not being used to hold the airplane up.

Another way of thinking about load factors is to consider them as measures of **centrifugal force.** We know that a load factor of greater than one will cause the airplane to assume a curved flight path in an upward direction. We could just as well think of the excess load factor as representing the centrifugal force required to keep the airplane in this curved path. If we tie a heavy object to a string and swing it in an arc, the string will have to pull harder than it did when the object was at rest. In aeronautics, the object is represented by the airplane and the pull of the string is replaced by the lift of the wing. Exactly the same physical laws and formulas apply in each case.

Connection between load factor and flight maneuvers

Now we can see that whenever the pilot causes the airplane to travel in a curved path he imposes an "excess" load factor on the wings. If his pull-up begins from level flight, he starts with one load factor and adds to that the extra load factors required to cause the curved flight path. But if he is diving vertically before the pull-up, the load factor he starts with is approximately zero, as the wing will no longer be lifting the airplane. Similarly, a perfectly circular loop done at constant speed would require a constant centrifugal force to hold the radius of curvature. Assume that this centrifugal force is represented by a load factor of 2. Then the pilot would have to apply a load factor of 3 at the beginning of the loop (airplane horizontal), 2 when going straight up or down on either side, and one when at the top (airplane inverted). This variation is, of course, due to the ever-present pull of gravity which acts away from the center of the circle at the botton, tangent to the circle at the sides, and toward the center when the airplane is at the top. (See Fig. 1.)

Load factors in steep turns

A banked turn is no exception to the above rule. If we "look down on top" of this maneuver we can see that the airplane is going in a circle and that there must be a "horizontal" force acting toward the center of this circle (representing the hypothetical "string"). Looking straight at the airplane (Fig. 1) we can see that this force is produced by the wing, which has been banked for this purpose. In a perfect bank the wing will have just the right slant to produce the required centrifugal force and also the vertical force required to overcome the pull of gravity, as shown in Fig. 1. In this figure, it is assumed that a load factor of 2 is required to hold the radius of the turn. The wing load factor must then be great enough to offset both the 2 (centrifugal) and the one (gravity). This works out to an actual load factor of 2.24 in this case, which represents an angle of bank of about 63 degrees.

The load factors required to hold a given angle of bank without slipping or "squashing" are given in table 1.

Table I reveals an interesting fact, which is that it requires an angle of bank of about 70 deg. to produce the load factor of 3 previously referred to as a safe value for any small airplane. This degree of bank is usually considered "vertical" bank and should be regarded as the upper limit for small commercial airplanes. Note that the load factor increases very rapidly as an angle of 70 deg. is exceeded.

The values given in Table I are based on the assumption that all the lift is derived from the wings. It is possible to make a true vertical bank without exceeding safe load-factor values if enough lift can be obtained from the fuselage and propeller pull to balance the pull of gravity. This is usually impossible with a small low-powered airplane. Pilots should therefore attempt to hold the angle of bank to less than 70 deg. in steep turns in order to avoid any possibility of exceeding a safe load factor.

Load factors in pull-ups

An abrupt pull-up at high speed is by far the most likely manuever to cause structural trouble. At any speed much greater than about twice the stalling speed, the pilot has it within his power to pull the wings off of almost any airplane. No matter how strong the designer makes the airplane, there will always be some speed above which the pilot can break the wings in an abrupt pull-up. The only exceptions to this would be when the airplane was so strong that the pilot "passed out" first (as in some military types) or when the designer deliberately incorporated some special means of limiting the pull-up. The latter method has not been used to any extent and there is probably no commercial airplane in this country that could safely permit the pilot to pull up to load factors at which he would "go black".

Practically all cases of structural failures in flight are caused by too abrupt pull-ups at high speeds. It is therefore important to know two things: (1) At what speed do abrupt pull-ups become dangerous and (2) how hard a pull-up can be made with safety at speeds above this value. The answer to the first question is relatively simple. As a rough approximation we can say that the maximum safe speed for abrupt pull-up is about twice the normal stalling speed ("normal" meaning at design gross weight). To be perfectly safe, it would probably be advisable to confine violent maneuvers to speeds even lower than this.

The limiting speed for abrupt maneuvers might, for convenience, be called the MANEUVERING speed. The reason why such a speed exists is that the wing will stall if the pilot tries to produce a high load factor at a relatively low speed. The stalling speed usually referred to is the stalling speed at a load factor of one, that is, level flight. But at higher load factors the stalling speed is also higher. The increase is proportional to the square root of the load factor. Thus if we raise the load factor from one to four, the stalling speed is increased by the square root of four, which of course means that the normal stalling speed is doubled. Thus if we are flying at twice the normal stalling speed we can theoretically pull-up to a load factor of 4, but no higher. If we tried to exceed 4, the wing would stall. This is what happens if the pilot tries to make a steep turn with insufficient speed: The wings will stall before the necessary load factor can be developed. In fact it is possible to use Table I to determine the minimum speed required for a given angle of bank, simply by taking the square root of the load factors given in that table. See Table II.

In Table II, a column has been added to indicate how the actual stalling speed will vary for an airplane having a normal stalling speed of 50 miles per hour. Note that a 70 deg. bank cannot be properly made at a speed below 85 miles per hour.

These examples are given to show how the load factor is directly connected with flight maneuvers of various kinds. Going back to the pull-up we can see that there will be no danger as long as the speed is held below the so-called MANEUVERING speed. Since this speed is actually the stalling speed corresponding to the maximum safe load factor, we can easily calculate it by taking the square root of that load factor. Thus for a load factor of 3, the MANEUVERING speed would be the square root of 3 or 1.73 times the normal stalling speed. If the latter were 40 miles per hour, the maneuvering speed would be 40 × 1.73 or about 69 miles per hour. As a rough guide, we can therefore say that **maneuvers involving sharp pull-ups (such as the snap roll) should be performed at speeds below about 70 miles per hour,** for the average small airplane. At higher speeds the pilot must depend on his physical sensations to tell him what load factor he is getting. We can now discuss what load factors mean to the pilot in terms of physical sensation.

Measuring load factors

Load factors can be measured by an instrument or estimated from their effect on the body. The instrument used is an accelerometer, which indicates load factor directly. This instrument sometimes has a maximum-reading needle, which will tell the pilot what maximum load factor he produced in a maneuver. Accelerometers are not in common use on small commercial airlanes, but they are valuable as a means of familiarizing pilots with load factors of various magnitude. If an accelerometer is used to a considerable extent during the training period, the pilot will tend to develop an instinctive feeling for load factors. In fact, the accelerometer can even be used as an aid in flight training, if the pilot understands the true significance of load factors. Racing pilots, for instance, have found the instrument useful in maintaining a constant radius of turn around a pylon.

If no accelerometer is available, the pilot must rely on his own good judgment and physical sensations, when making pull-ups at speeds greater than the MANEUVERING SPEED. It should be noted that the "feel" of a load factor will depend very much on the **time of application.** Pilots can withstand load factors as high as 9 or 10 if they are only briefly applied. But if the load factor is sustained for several seconds it is difficult to stand more than 4 or 5. At any rate, pilots of small commercial airplanes evidently cannot expect to pull up to the maximum load factor they can stand, if they want the wings to stay on. Unless a special airplane is used (such as a military or acrobatic type) the pilot will have to try to stay below a load factor of about 3, or 4 at the most, when making pull ups.

It may be of interest to realize that a load factor of 3 can be obtained in any type of swing or pendulum device, simply by starting from a horizontal position. In this case the factor of 3 acts only at the very bottom of the swing. If the start is made from the "top" (as in the thrill rides often found at amusement parks) a load factor of 5 is developed at the bottom of the circle.

Flight maneuvers

Practically any flight maneuver can be accomplished by a skillful pilot without exceeding a load factor of 3. The only rule necessary to remember is to make pull-ups gradually. If a sharp pull-up is desired, as in a snap roll, the speed should first be reduced to not more than approximately 70 miles per hour depending on the airplane in question. A **snap roll** from high speed or a steep dive is almost sure to cause excessive load factors, as the maneuver necessitates stalling the wing, which in turn requires a high load factor at high speeds.

Acro II biplane is a rugged design capable of sportsman and intermediate aerobatics. Design will accept powerplants from 108 hp to 200 hp.

Normal loops are not dangerous if properly performed. Here again the rule to make pull-ups gradually will avoid undue loads on the structure. Low-powered airplanes are likely to be subjected to higher load factors in loops, due to the necessity of diving to a high speed and then performing the maneuver rapidly. Likewise, the student pilot often has the tendency to pull out of the loop too rapidly after its completion.

Tail slides often result from incompleted loops of whip stalls and can cause considerable damage to the structure. **Airplanes are not designed to be flown backwards** and damage of ribs or rear lift bracing is almost sure to result from a tail slide of 100 feet or more. Proper instruction should eliminate the possibility of a bad tail slide, as the airplane can easily be righted before gaining any appreciable speed.

Spins are not dangerous structurally, but proper precautions should be taken against abrupt pull-ups in recovering from a spin, as the speed is likely to be above the safe "maneuvering" speed.

Dives should not be made to excessive speeds, as this increases the danger of making too severe a pull out. At very high diving speeds there is also more danger of flutter. The "NEVER EXCEED" speed on the placard of every certificated airplane represents the absolute upper limit for which the airplane has been analyzed and tested. Although the airplane may be good for higher speeds, its airworthiness at such speeds has not been established and is therefore highly questionable. Exceeding the maximum placard speed is therefore to be considered as a violation of the principles of safe operation. At any speed in this region, in fact, the pilot will have to be extremely careful, as it is an easy matter to produce a dangerous load factor by a slight movement of the elevator control.

Gusts. All certificated airplanes are designed to take the loads imposed by gusts of considerable intensity. Gust load factors increase with increasing air speed and the value used in design usually corresponds to the high speed in level flight. In extremely rough air the safest procedure is to reduce the speed to the "maneuvering" speed, as it is then impossible for gusts to produce dangerous load factors. Several accidents have occurred due to diving to high speeds in rough ar. As a general rule, **the rougher the weather, the slower the airplane should be flown.**

Weight. Any airplane designed for a certain load factor at gross weight can safely withstand higher load factors at reduced weight. All of the foregoing rules are based on the assumption that the airplane is loaded to the maximum gross weight. If overloaded, the allowable load factors will be reduced accordingly and the pilot is likely to damage the structure in maneuvers that would normally be quite safe.

On the other hand, if the weight can be reduced 25 percent, for example, the allowable load factors are correspondingly increased. The maximum recommended value of 3 would then become 4 (although the "maneuvering speed" would remain about the same). Two rules can be derived from this fact:

1) Never overload the airplane.

2) Always fly as light as possible when performing acrobatics.

The second rule may explain why many light airplanes seem to be capable of withstanding very severe maneuvers. When lightly loaded, the pilot can "get away" with a lot of stunts that would be decidedly dangerous with a full load. Therefore, don't load your friends in your aircraft, with a full gas load, some baggage and parachutes all around and proceed to put the airplane through its paces. If you do, the parachutes will probably be necessary.

Summary

(a) Become familiar with load factors; (b) don't make abrupt pull-ups at high speeds; (c) keep the speed below 70 miles per hour for snap rolls and abrupt pull-ups; (d) don't let the airplane make "tail slides"; (e) slow down in rough air; (f) fly light when doing acrobatics; (g) don't make banks over 70 degrees; (h) never attempt inverted loops.

Pober Pixie nearing completion. Note clecos in engine cowl, wings positioned on fuselage and held in place by wood braces prior to fabrication of lift struts. Aluminum leading and trailing edges, as well as aileron fittings have yet to be installed on the wing.

Nothing more rewarding than displaying your completed aircraft at a local or regional fly-in. Best yet is attending aviation's greatest event — EAA Oshkosh! View below shows just a portion of aircraft that attended the 1990 event. Homebuilt aircraft are parked from the taxiway where the transport aircraft can be seen (including the supersonic Concorde), all the way left to the end of the main runway. Hundreds and hundreds of homebuilts of incredible workmanship and beauty!

PART III

EAA Safety Check List

Spend some time with your plane and this check list before those first flights. Thirty minutes with a pencil here may be worth the rest of your life.

PROPELLER	Yes	No
1. Blades		
Laminations not separated?		
Breaks, scratches, nicks, tipping?		
Loose rivets in tipping?		
Drain holes in tip clear?		
2. Hub		
Any cracks or corrosion?		
Hub properly seated and safetied?		
3. Control Mechanism		
Oil leaks?		
Worn bearings?		
Secure?		
4. Attachment		
All bolt & nut threads undamaged?		
All bolts & nuts secured & safetied?		
5. Spinner		
Cracks?		
Properly secured?		
Is spinner chafing into prop?		

ENGINE & ENGINE COMPARTMENT	Yes	No
1. Fuel System		
All lines of approved type?		
All strainers clean?		
All lines secured against vibration?		
Gascolator bowl at low point in system when aircraft is in normal ground position?		
Fuel drains operative?		
All connections properly tightened?		
2. Oil System		
All lines of approved type?		
All lines secured against vibration?		
Oil tank has no cracks or leaks?		
Tank properly secured and safetied?		
All plugs and strainers cleaned and safetied?		
3. Ignition—Electrical System		
All wiring proper type and gauge?		
All fastenings secured and safetied?		
Magnetos properly grounded?		
Spark plugs cleaned & undamaged?		
Spark plugs properly torqued?		
Engine grounded to airframe?		
Starter/generator secured?		
4. Exhaust Manifold		
Secured and safetied?		
All gaskets in good condition?		

ENGINE & ENGINE COMP'T.	Yes	No
All stacks in good condition — no cracks or rusted-out areas?		
Carb heat and cabin heat muffs removed and manifold inspected?		
5. Controls		
All secured and safetied?		
No excessive play in any linkages?		
No interference between any control and the structure throughout the full operating range?		
Carb heater gate open & close fully?		
6. Mount		
Secured and safetied?		
All joints inspected for cracks?		
Any bends in mount tubes?		
Bushings in good condition?		
7. Cowlings		
Secured and/or safetied?		
All latches or fastenings working properly?		
Any cracks properly checked or reinforced?		
Cowlings clean?		
8. Power Plant in General		
All necessary safeties, palnuts, locknuts, etc. in place?		
No fuel or oil leaks?		
All accessories secured & safetied?		

FUSELAGE — HULL	Yes	No
1. Structure		
All welds sound?		
All tubing straight and uncracked?		
No rust or corrosion?		
All attach fittings sound, no cracks, elongation of holes or worn threads?		
All rivets properly installed?		
Inspection openings for all vital areas?		
Fuselage properly drained, that is, no built-in moisture traps?		
Firewall of proper fireproof material?		
2. Cover		
Properly attached?		
No tears, distortions, or abrasions?		
Any breaks or ruptures properly repaired?		
3. Control System		
Properly secured and safetied?		
Controls stops provided & adjusted?		
All fittings of proper thread & size?		

FUSELAGE — HULL (Continued)	Yes	No
All pulleys of proper diameter for bends, proper size for cable, and guarded?		
All cable of proper size (1/8" min.) and condition?		
Any parts in system subject to rotation for any reason properly secured and safetied?		
Return springs on rudder pedals?		
No interference between any control part (cable, tube, or linkage) and any other part of the structure throughout full control movement?		
Adequate room for full control throw when aircraft is occupied?		
Controls arranged to minimize danger of blocking by foreign objects?		
Grip properly secured to control stick or wheel?		
4. Electrical System		
All grommets, particularly in firewall, snug fitting and in good condition?		
All wires of proper gauge, insulated, and secured?		
Wires do not rest on abrasive surfaces?		
Battery installation of sufficient strength?		
Battery properly ventilated and drained?		
No corrosion at or around battery or its vents?		
Fuses of adequate amperage?		
5. Fuel System — Tanks		
Drains properly located to discharge clear of aircraft?		
All outlets properly screened?		
Breather inlets clear?		
Fuel shut-off valve installed?		
Fuel shut-off valve easily reached by pilot?		
All fuel lines of proper approved type?		
All fuel lines secured against vibration?		
Is tank located so that sufficient head is available in maximum climb with minimum fuel? Placard if necessary?		
Has tank sufficient expansion area?		
Any tank overflow discharge clear of hazardous areas on aircraft?		
Is tank support sufficient to meet strength requirements?		
Does tank clear surrounding structure?		
Do tank supports minimize strain and chafing?		

To insure its safe construction and operation, and to further emphasize the vital necessity for thorough consideration of every item which goes into your airplane, the following working check-list should be used, and it is suggested that it be made a part of the aircraft records.

EXITS	Yes	No
1. Can aircraft be cleared rapidly in case of emergency?		
Are special precautions available during test period, such as jettisonable doors or canopy?		
If parachute is to be worn, does it clear all controls?		
Baggage Compartment		
1. Are walls and floors of sufficient strength to withstand flight loads?		
Can anything escape from baggage compartment by accident?		
Cabin—Cockpit		
1. Instruments		
Are all instruments functioning and accurate?		
Are all instruments marked, max pressures, temperatures, speeds?		
Are all vital instruments easily visible to pilot?		
2. Flight—Engine Controls		
Are all engine controls marked or easily identifiable?		
Are all engine controls smooth in operation, without excessive resistance, & easily available to pilot?		
Are all flight controls arranged so that jamming by dropped gloves, etc. is impossible?		
3. Fuel Systems		
Are all gas valves easily reached by pilot?		
Are all gas valves marked ON, OFF, LEFT, RIGHT?		
Are all gas valves in such a position that accidental operation is impossible, or guarded in such a way that accidental operation is impossible?		
4. Seats		
Are seats of sufficient strength for maximum flight loads contemplated?		
Does seat "flex" enough at any time to interfere with flight controls?		
5. Safety Belts and Shoulder Harness		
Is installation and attachments of sufficient strength to meet 9G forward load minimum?		
Does attachment connect directly to primary structure?		
Are belts and harness in top condition?		

EXITS (Continued)	Yes	No
Is belt of correct size, that is, no long over-tongue?		
Is a separate belt and shoulder harness supplied for each occupant?		
6. Heating—Ventilation		
Is cabin or cockpit in negative pressure area and liable to suck in exhaust fumes?		
Is any provision made for ventilating cabin other than normal leakage?		
7. Windshield—Windows		
Are windshield and windows of recognized aeronautical materials?		
Is windshield braced against positive or negative pressures in flight, either by design or extra bracing?		
WING — TAIL SURFACES		
Fixed Surfaces		
Are all interior fastenings secured and/or safetied?		
Is interior properly weatherproofed?		
Have any mice been inside lately?		
Movable Surfaces		
Are stops provided, either at wing or somewhere else in the control system?		
Are all hinges and brackets sound?		
Are all hinge pins secured and safetied?		
Is there any excessive play in hinges?		
Is there any excessive play in control cables or tubes?		
External Bracing		
Is the interior of all struts weather protected?		
Are all adjustable fittings locked, secured and safetied?		
Are struts undamaged by bends or dents?		
Are all wires serviceable with proper end fittings?		
Attach Fittings		
Are bolts of proper size installed?		
Are all bolts secured and safetied?		
Have all bolts been examined for wear?		
Flight Control Mechanism		
All cables and tubes unbroken or unbent & with proper end fittings?		
All control attachments secured and safetied?		

WING—TAIL SURFACES (Cont.)	Yes	No
All pulleys free from interference and guarded?		
All torque tubes and bell cranks in good condition?		
No interference with fuselage or wing structure throughout full control travel?		
Fuel Tanks		
See Fuselage Section Also		
Are drains supplied at low point in tank when aircraft is in normal ground position?		
Fuel overflow drains clear of aircraft—no tendency for overflow to soak into aircraft structure?		
LANDING GEAR		
Properly lubricated?		
Proper oleo inflation?		
Shock cords or springs in good condition?		
All attach fittings uncracked and sound?		
All bolts holes not elongated?		
All attach bolts secured & safetied?		
Brake lines in good condition?		
Brakes operating properly?		
Correct hydraulic fluid in lines?		
Wheels uncracked?		
Tires unworn & properly inflated?		
Excessive side play in wheel bearings?		

GENERAL

ALL BOLTS, WHEREVER POSSIBLE, HEAD UP AND FORWARD.

All exterior fastenings visible from cockpit or cabin should have safetied end toward pilot, wherever possible.

A complete walkaround inspection of the aircraft should be accomplished to check that every bolt visible on the exterior is secured and safetied. That there is no visible structural damage. That all inspection panels and covers are in place and attached. That all parts of the aircraft are in proper alignment.

DON'T FORGET TO PUT IN ENOUGH GAS PRIOR TO THAT FIRST FLIGHT — GROUND RUNNING AND TAXI TESTS CAN USE UP A LOT MORE THAN YOU THINK!

OK — Kick the tires, add another coat of paint and AWAY WE GO.

Top — Wittman Tailwind was one of the earliest high performance aircraft available to homebuilders. Designed by famed speed pilot Steve Wittman, the Tailwind's outstanding performance still holds its own against today's latest composite designs. Aircraft shown built by Kay Downs, Costa Mesa, CA.

Above — What could be finer than a homebuilt on floats? This Kitfox Model III sports a set of Aeroset composite floats. Powerplant is a Rotax 912, a 4-stroke 80 hp engine.

Above-Left — Replicas of World War I aircraft are quite popular with EAAers. This Fokker Triplane was built by Dick Curtis, Wichita, KS. It is powered with a Continental W-670 of 220 hp.

Left — Steen Skybolt built by S.H. "Wes" Schmid and Clarence Schrieber, Milwaukee, WI over an eight year period. Engine is a 180 hp Lycoming O-360 A4A. Full inverted fuel system is installed.

The Annual and 100-Hour Inspection

Part 43, App. D

APPENDIX D — Scope and Detail of Items (as applicable to the particular aircraft) to be included in annual and 100-hour inspections.

(a) Each person performing an annual or 100-hour inspection shall, before that inspection, remove or open all necessary inspection plates, access doors, fairing, and cowling. He shall thoroughly clean the aircraft and aircraft engine.

(b) Each person performing an annual or 100-hour inspection shall inspect (where applicable) the following components of the fuselage and hull group:

(1) Fabric and skin — for deterioration, distortion, other evidence of failure, and defective or insecure attachment of fittings.

(2) Systems and components — for improper installation, apparent defects, and unsatfisfactory operation.

(3) Envelope, gas bags, ballast tanks, and related parts — for poor condition.

(c) Each person performing an annual or 100-hour inspection shall inspect (where applicable) the following components of the cabin and cockpit group:

(1) Generally — for uncleanliness and loose equipment that might foul the controls.

(2) Seats and safety belts — for poor condition and apparent defects.

(3) Windows and windshields — for deterioration and breakage.

(4) Instruments — for poor condition, mounting, marking, and (where practicable) improper operation.

(5) Flight and engine controls — for improper installation and improper operation.

(6) Batteries — for improper installation and improper charge.

(7) All systems — for improper installation, poor general condition, apparent and obvious defects, and insecurity of attachment.

(d) Each person performing an annual or 100-hour inspection shall inspect (where applicable) components of the engine and nacelle group as follows:

(1) Engine secttion — for visual evidence of excessive oil, fuel, or hydraulic leaks, and sources of such leaks.

(2) Studs and nuts — for improper torquing and obvious defects.

(3) Internal engine — for cylinder compression and for metal particles or foreign matter on screens and sump drain plugs. If there is weak cylinder compression, for improper internal condition and improper internal tolerances.

(4) Engine mount — for cracks, looseness of mounting, and looseness of engine to mount.

(5) Flexible vibration dampeners — for poor condition and deterioration.

(6) Engine controls — for defects, improper travel, and improper safetying.

(7) Lines, hoses, and clamps — for leaks, improper condition and looseness.

(8) Exhaust stacks — for cracks, defects, and improper attachment.

(9) Accessories — for apparent defects in security of mounting.

(10) All systems — for improper installation, poor general condition, defects, and insecure attachment.

(11) Cowling — for cracks, and defects.

(e) Each person performing an annual or 100-hour inspection shall inspect (where applicable) the following components of the landing gear group:

(1) All units — for poor condition and insecurity of attachment.

(2) Shock absorbing devices — for improper oleo fluid level.

(3) Linkages, trusses, and members — for undue or excessive wear fatigue, and distortion.

(4) Retracting and locking mechanism — for improper operation.

(5) Hydraulic lines — for leakage.

(6) Electrical system — for chafing and improper operation of switches.

(7) Wheels — for cracks, defects, and condition of bearings.

(8) Tires — for wear and cuts.

(9) Brakes — for improper adjustment.

(10) Floats and skis — for insecure attachment and obvious or apparent defects.

(f) Each person performing an annual or 100-hour inspection shall inspect (where applicable) all components of the wing and center section assembly for poor general condition, fabric or skin deterioration, distortion, evidence of failure, and insecurity of attachment.

(g) Each person performing an annual or 100-hour inspection shall inspect (where applicable) all components and systems that make up the complete empennage assembly for poor general condition, fabric or skin deterioration, distortion, evidence of failure, insecure attachment, improper component installation, and improper component operation.

(h) Each person performing an annual or 100-hour inspection shall inspect (where applicable) the following components of the propeller group:

(1) Propeller assembly — for cracks, nicks, binds, and oil leakage.

(2) Bolts — for improper torquing and lack of safetying.

(3) Anti-icing devices — for improper operations and obvious defects.

(4) Control mechanisms — for improper operation. Insecure mounting, and restricted travel.

(i) Each person performing an annual or 100-hour inspection shall inspect (where applicable) the following components of the radio group:

(1) Radio and electronic equipment — for improper installation and insecure mounting.

(2) Wiring and conduits — for improper routing, insecure mounting, and obvious defects.

(3) Bonding and shielding — for improper installation and poor condition.

(4) Antenna including trailing antenna — for poor condition, insecure mounting, and improper operation.

(j) Each person performing an annual or 100-hour inspection shall inspect (where applicable) each installed miscellaneous item that is not otherwise covered by this listing for improper installation and improper operation.

1991 EAA International Chapter Directory

BUENOS AIRES
722 BUENOS AIRES, ARGENTINA
Ildefonso D. Durana
Valle 1362
1406 Buenos Aires
Argentina
TELEPHONE: 432-1330
MEETING: 1ST WEDNESDAY, 8:00 PM
AUTODROMO PUERTA 9

CZECHOSLOVAKI
955 CZECHOSLOVAKIA, BRNO
Petr Dobias
Matulkova 4
Brno 61200
Czechoslovakia
TELEPHONE: 05-749280
MEETING: CONTACT PRESIDENT

DENMARK
655 DENMARK, SPJALD
Magnus Pedersen
Hovedgaden 54
DK 6971 Spjald
Denmark
TELEPHONE: 97-381020
MEETING: CONTACT PRESIDENT

INDIA
962 INDIA, MADRAS
Kanagaraj Subramanian
N06 Sapthagiri Kannan Illam
Tngos Colony Adambakkam Madras
600088
India
MEETING: 1ST SUNDAY, 3:00 PM
CONTACT PRESIDENT

ITALY
459 ITALY, VENEGONO SUPERIORE
Mario Puppi
Via Piceni 1
23100 Gallarate VA
Italy
TELEPHONE: 0331-782-582
MEETING: 1ST SATURDAY, 9:00 PM
CONTACT SECY 0331-865-204

KOREA
930 KOREA, SEOUL
Won Bok Lee
1-1209 Daegyo Apt Yoido-Dong
Young Deumgpo-Ku
150-010 Seoul
Korea
TELEPHONE: 02-785-4158
MEETING: 3RD FRIDAY, 1:00 PM
EVERY OTHER MONTH
JINDO BLDG, #1301, 168-5
DOWHA-DONG, MAPO-KU

MEXICO
815 MEXICO, MEXICO CITY
Marcial Lara Sanchez
Urales 42 4A Secc Lomas Verdes
Naucalpan Edo 9G
Mexico 53120
TELEPHONE: 905-393-3056
MEETING: CONTACT PRESIDENT

NETHERLANDS
664 NETHERLANDS, MAASBOMMEL
Wolanda Hornstra
Zebraspoor 698
3605 HP Maarssen
Netherlands
TELEPHONE: 31-3465-68653
MEETING: LAST SATURDAY, 10:00 AM
CONTACT PRESIDENT

NEW ZEALAND
666 NEW ZEALAND, AUCKLAND
Allan R. McCreadie
P.O. Box 97
Manurewa
New Zealand
TELEPHONE: 64 9-2673410
MEETING: LAST FRIDAY, 7:30 PM
ST. HILDA'S CHURCH HALL
PENROSE ROAD

NORWAY
573 NORWAY, OSLO
Knut Brdreskift
Hennumbratan 22
3408 Tranby
Norway
TELEPHONE: 03-851995
MEETING: CONTACT PRESIDENT

SO AFRICA .
592 SO AFRICA, CAPE TOWN
Andres W. Van Dikjk
5 Danie Uys Str
Stikland Ind Bellville 7530
South Africa
TELEPHONE: 021-965-359
MEETING: LAST SATURDAY, 3:00 PM
FISANTEKRAAL AIRFIELD

645 SO AFRICA, DURBAN
Graham J. McDonald
16 Glen Drive
Kloof 3610
South Africa
TELEPHONE: 031-764-3826
MEETING: CONTACT PRESIDENT
VIRGINIA AIRPORT

SO AFRICA
357 SO AFRICA, PIETERMARITZBG
Ian R. Drennan
24 Christie Road
Pietermaritzburg 3201
TELEPHONE: 0331-63063
MEETING: 1ST FRIDAY, 7:30 PM
PIETERMARITZBURG AERO CLB

778 SO AFRICA, PORT ELIZABETH
Bruce Turner
62 6th Avenue Newton Park
Port Elizabeth 6045
South Africa
TELEPHONE: 041-353333
MEETING: 1ST SATURDAY, 2:00 PM
UTTENHAGE AIRFIELD

322 SO AFRICA, TRANSVAAL
Peter O. Hengst
P.O. Box 195
Bergvlei
South Africa
TELEPHONE: 011-805-3837
MEETING: 2ND WEDNESDAY, 7:30 PM
CIVIC HALL, HALFWAY HOUSE

SPAIN
899 SPAIN, CACERES
Joaquin Guerra Jimenez
Angel Fernandez, 24
10004 Caceres
Spain
TELEPHONE: 927-229469
MEETING: CONTACT PRESIDENT

SWEDEN
222 SWEDEN, STOCKHOLM
Curt Sandberg
Astrakanvagen 6
S-17800 Ekero
Sweden
TELEPHONE: 0756 320 62
MEETING: CONTACT PRESIDENT
BARKARBY AIRFIELD

Above — 6Y-JHD, a Pitts S1 is the result of the Jamaican Flying Club started by two World War II RAF veterans with a minimum of finances but a wealth of aeronautical knowledge. Forming a club to share expenses the Pitts was started and completed in 1971. It has flown regularly since. Club members continue to build aircraft!

Left — The Maverick is a new Australian design by Graham Swannell. It is single place composite, fully aerobatic with a 150 knot cruise speed. Power is a Continental 0-200 fitted with an Ellison Throttle Body Injector. Aircraft was built by Geoff Danes who flies Jetstreams and HS 125s for a living.

Below — About to touch down on the River Swan, Perth, Australia is John Brown's all-wood Volmer Sportsman amphibian. Aircraft was part of a fly-in held in Perth in which over 30 types of aircraft put in an appearance. Flying and homebuilding are very popular throughout Australia and New Zealand.

Above — Erland Magnusson of Sweden flew this Thorp T-18 in 1989. It was under construction over a ten year period. This aircraft represents the second Thorp T-18 built in Sweden. The side by side, all-metal aircraft is powered with a Lycoming O-320 E2A mounting a 68 x 74 propeller.

"How To" EAA Videos to Help You Build

When it comes to doing it yourself, there is a right way . . . and a wrong way.
With help from EAA's "How to" Videotape Series, you can keep your homebuilt project moving in the right direction . . .

BASIC AIRCRAFT PAINTING Join a professional paint crew as they show you the detailed steps and techniques involved in painting an all metal aircraft. (60 min.)
VHS 21-36467 **BETA 21-36468**

BASIC AIRCRAFT WELDING
Learn the intricacies of welding with practical demonstrations. An excellent video for the homebuilder. (30 min.)
VHS 21-35773 **BETA 21-35769**

BASIC AIRCRAFT WOODWORKING
Woodworking knowledge is essential to any homebuilding project. Power tool safety is also discussed. A great starter tape. (30 min.)
VHS 21-35776 **BETA 21-35775**

BASIC AIRCRAFT WOODWORKING AND WELDING
Both programs on one cassette.
VHS 21-35789 **BETA 21-35788**

FABRIC COVERING WITH RAY STITS
Learn the delicate art of fabric covering from the best — Ray Stits (EAA #136) — the man who developed the Stits Poly-Fiber Aircraft Coating process. Step-by-step instructions are detailed in this excellent video. (120 min.)
VHS 21-36141 **BETA 21-36142**

VW AIRCRAFT ENGINE ASSEMBLY
How to take a stock VW "Bug" engine and convert it for aircraft use. Includes step-by-step graphic manual. (120 min.)
VHS 21-36371

VW AIRCRAFT INSTALLATION
Shows how to mount the VW "Bug" engine in a typical homebuilt. (60 min.)
VHS 21-36369

THE COMPLETE RUTAN ENCYCLOPEDIA FOR AIRCRAFT HOMEBUILDERS
See Burt Rutan's VariEze, Long-EZ and Defiant in development, construction and in the air. (70 min.)
VHS 21-35832 **BETA 21-35833**

BASIC COMPOSITE CONSTRUCTION
Learn the basics of aircraft fiberglass construction. Resins, hot wire cutting techniques, tips and more. (45 min.)
VHS 21-36029 **BETA 21-36028**

BUILDING THE RUTAN COMPOSITES
A step-by-step program featuring Burt Rutan showing the complete process for utilizing his fiberglass-epoxy-foam construction. 91 min.)
VHS 21-35830 **BETA 21-35831**

15% DISCOUNT
TO EAA VIDEO CLUB MEMBERS
(call for info and details)

BEDE ON DESIGN
Renowned aircraft designer and EAA member Jim Bede relates his thoughts and findings on aerodynamics, design, airfoils, control surfaces and more. These nine programs may be purchased separately or in two sets.

SET A: (programs 1-7) Aircraft design.
VHS 21-35989 **BETA 21-35999**

SET B: (programs 1-9) Aircraft design and aerodynamics.
VHS 21-36020 **BETA 21-36021**

BEDE ON DESIGN
*TAPE ONE: Understanding Aerodynamic Drag (59 min.)
VHS 21-35980 **BETA 21-35990**

*TAPE TWO: Selecting the Best Airfoil (41 min.)
VHS 21-35981 **BETA 21-35991**

*TAPE THREE: Aircraft Wing Design (35 min.)
VHS 21-35982 **BETA 21-35992**

*TAPE FOUR: High Lift Devices for Aircraft (33 min.)
VHS 21-35983 **BETA 21-35993**

*TAPE FIVE: Understanding Airplane Stability (46 min.)
VHS 21-35984 **BETA 21-35994**

*TAPE SIX: Aircraft Control Surfaces, Design (73 min.)
VHS 21-35985 **BETA 21-35995**

*TAPE SEVEN: Design Procedures for Engine Cooling (44 min.)
VHS 21-35986 **BETA 21-35996**

*TAPE EIGHT: What Pilots Should Know about Aerodynamics (38 min.)
VHS 21-35987 **BETA 21-35997**

*TAPE NINE: Understanding the Aerodynamics of Automobiles (46 min.)
VHS 21-35988 **BETA 21-35998**

CALL TOLL FREE 1-800-843-3612
Outside USA Call (414) 426-4800
Fax Orders Accepted (414) 426-4873

EAA Manuals And Publications

Aircraft Welding (94 pgs) *21-37864*
Fundamental welding techniques for the building & repair of aircraft, from the pages of Sport Aviation and other sources. 94 pages filled with aircraft welding tips and information.

Wood Aircraft Building Techniques (140 pgs) *21-18100*
Excellent resource book on "How To" build or repair wood aircraft.

Custom Built Sport Aircraft Handbook (144 pgs) *21-13510*
A guide to construction standards for the amateur aircraft builder and detailed information on FAA and Canadian contact and applicable FARs.

CAM-18 — Aircraft Maintenance Manual (209 pgs) *21-13460*
Easy to read guide to maintenance, repair & alteration of aircraft. Includes airframes, powerplants, propellers & appliances. CAM 18 is the early version of AC-43.13, current guideline to repair of type certificated aircraft.

CAM-107 — Aircraft Powerplant Handbook (350 pgs) *21-13470*
C.A.A. Technical Manual for Maintenance and Repair of Aircraft Powerplants.

Firewall Foreward (295 pgs) *21-13950*
By Tony Bingelis. Manual on piston engine installations. One of the best engine reference manuals for the amateur builder.

Sportplane Builder (300 pgs) *21-30140*
By Tony Bingelis. (Vol. I) Aircraft construction methods and techniques for the homebuilder. Articles taken from Tony's columns in Sport Aviation magazine.

Sportplane Construction Techniques (360 pgs) *21-01395*
By Tony Bingelis. (Vol. II) More aircraft construction tips for the homebuilder. Articles taken from Tony's columns in Sport Aviation magazine.

Flying and Glider Manuals
Reprints of original "Building and Flying Manuals" published from 1929 to 1933 by Modern Mechanix and Inventions.

1929 Flying & Glider Manual *21-14167*
Contains information on flight lessons plus building the Heath Super Parasol, Russell-Henderson Light Monoplane and an easy to build glider.

1930 Flying & Glider Manual *21-14168*
Plans for building a Heath Baby Bullet, set of light plane metal floats; building the Northop Glider, Lincoln Biplane, Alco Sportplane, plus other tips on building and welding.

1931 Flying & Glider Manual *21-14169*
Building the "Longster", Georgias Special, a glider and secondary glider, Driggs Dart, the Church Midwing, the Heath Seaplane Parasol and its pontoons, the Northop Glider and other gliders.

1932 Flying & Glider Manual *21-14170*
Building the Pietenpol Aircamper with Ford motor conversion, Powell "P-H" Racer, the Heath Super Soar Glider, Penquin practice plane, Ramsey "flying bathtub" and other kids. Also, build your own hangar.

1933 Flying & Glider Manual *21-14171*
Building the Gere Sport Biplane, Pietenpol floats, Pietenpol Sky Scout, and Henderson Longster. Also, Long Harlequin Motor plans, a hydroglider and information on building propellers.

NOTES: